Creative Engagements with Ecologies of Place

This book explores an exciting range of creative engagements with ecologies of place, using geopoetics, deep mapping and slow residency to propose broadly based collaborations in a form of 'disciplinary agnosticism'.

Providing a radical alternative to current notions of interdisciplinarity, this book demonstrates the breadth of new creative approaches and attitudes that now challenge assumptions of the solitary genius and a culture of 'possessive individualism'. Drawing upon a multiplicity of perspectives, the book builds on a variety of differing creative approaches, contrasting ways in which both visual art and the concept of the artist are shifting through engagement with ecologies of place. Through examples of specific established practices in the UK, Australia and the USA, and other emergent practices from across the world, it provides the reader with a rich illustration of the ways in which ensemble creative undertakings are reactivating art's relationship with place and transforming the role of the artist.

This book will be of interest to artists, art educators, environmental activists, cultural geographers, place-based philosophers and postgraduate students and to all those concerned with the revival of place through creative work in the twenty-first century.

Mary Modeen, Associate Dean (International) and Chair of Interdisciplinary Art Practice, researches broadly across Art and Humanities at the University of Dundee. She explores perception and place-based research, connecting many concerns. As such, this research is usually interdisciplinary, appearing as both art and writing.

Iain Biggs is a former Director of the PLaCE research centre at the University of the West of England (UWE), Bristol, an Honorary Research Fellow at the University of Dundee, and a Visiting Research Fellow at the Environmental Humanities Research Centre at Bath Spa University.

Routledge Research in Culture, Space and Identity

Series editor: Dr. Jon Anderson, School of Planning and Geography, Cardiff University, UK

The *Routledge Research in Culture, Space and Identity Series* offers a forum for original and innovative research within cultural geography and connected fields. Titles within the series are empirically and theoretically informed and explore a range of dynamic and captivating topics. This series provides a forum for cutting edge research and new theoretical perspectives that reflect the wealth of research currently being undertaken. This series is aimed at upper-level undergraduates, research students and academics, appealing to geographers as well as the broader social sciences, arts and humanities.

Artistic Approaches to Cultural Mapping
Activating Imaginaries and Means of Knowing
Edited by Nancy Duxbury, W.F. Garrett-Petts and Alys Longley

Geopoetics in Practice
Edited by Eric Magrane, Linda Russo, Sarah de Leeuw and Craig Santos Perez

Space, Taste and Affect
Atmospheres That Shape the Way We Eat
Edited by Emily Falconer

Geography, Art, Research
Artistic Research in the GeoHumanities
Harriet Hawkins

Creative Engagements with Ecologies of Place
Geopoetics, Deep Mapping and Slow Residencies
Mary Modeen and Iain Biggs

For more information about this series, please visit: www.routledge.com/Routledge-Research-in-Culture-Space-and-Identity/book-series/CSI

Creative Engagements with Ecologies of Place

Geopoetics, Deep Mapping and Slow Residencies

Mary Modeen and Iain Biggs

Routledge
Taylor & Francis Group

LONDON AND NEW YORK

First published 2021
by Routledge
2 Park Square, Milton Park, Abingdon, Oxon OX14 4RN

and by Routledge
52 Vanderbilt Avenue, New York, NY 10017

Routledge is an imprint of the Taylor & Francis Group, an informa business

British Library Cataloguing-in-Publication Data
A catalogue record for this book is available from the British Library

Library of Congress Cataloging-in-Publication Data
A catalog record has been requested for this book

ISBN: 978-0-367-54575-8 (hbk)
ISBN: 978-1-003-08977-3 (ebk)

Typeset in Times New Roman
by codeMantra

This book is respectfully dedicated to the memory of the polymath Tim Robinson: artist, cartographer, mathematician, fiction and non-fiction writer. His life and work embodied the engagement with, and wonder at, the richness and complexities of place that we hope to encourage here. This book is also dedicated to all who engage in creative engagements with ecologies of place: geo-poets, deep-mappers, artists who conduct slow residencies and all who conduct active practices to rebalance and restore our collective relationships with the earth we inhabit.

Contents

Figures

Acknowledgements

With so many people working in this area, the resources, research and conversation informing this book have been rich and varied. First and foremost, we have drawn on sharing working experience with people with similar ensemble concerns. Exchanges with Rowan O'Neill, Marlene Creates, Pauline O'Connell, Deirdre O'Mahony, Ciara Healy, Barbara Hawkins, Antony Lyons, Brett Wilson, Simon Read, Luci Gorrel Barnes, Mona Smith, Rebecca Krinke, Ruth Jones, Mike Pearson, Dominic Smith, Reiko Goto and Tim Collins, along with others far too numerous to mention, have been invaluable in providing testing grounds for our thinking. Then there are theorists in philosophy, environmental studies, human and cultural geography, anthropology, art and design, history and literature who have all influenced us, as have poets and visual artists concerned with the issues these theories address. The writings of Guattari, Foucault, Deleuze, Irigaray, Arendt, Bennet, Barad, Haraway, Tuan, Soja and a host of other theorists and philosophers who cover political, spatial and cultural issues have been important points of reference. We have derived particular support from socially engaged researchers who interact with site-specific selected areas and local communities. Artists, poets, designers and landscape architects – particularly those who clearly acknowledge the sense of a compound practice indicated above – have contributed to this work as they practice in environmentally engaged activities, as have our colleagues, past and present, and the innumerable students we have worked with. Those who have contributed photographs and permissions have made this book richer in content, especially Christine Baeumler, Gini Lee and Alexander and Susan Maris. We offer heart-felt thanks to them, to Nel Whiting and Katie Potapoff for editorial assistance, and to Faye Leerink and all those at Routledge who have made the publication of this book possible.

About the authors

Iain Biggs is the former Director of the PLaCE Research Centre at the University of the West of England, Bristol, and is currently an Honorary Research Fellow at the University of Dundee, Scotland, and Visiting Research Fellow at the Environmental Humanities Research Centre at Bath Spa University, England. With twenty years of experience in deep mapping, he currently contributes to various ecosophically oriented projects in the UK and Ireland. His chapter 'Ensemble Practices', for the ecology section of *The Routledge Companion to Art in the Public Domain*, will appear in 2020.

Mary Modeen is the Chair of Interdisciplinary Art Practices at the University of Dundee, Scotland, and Associate Dean International. She coordinates PhD studies and has founded and directed several interdisciplinary courses at the undergraduate and postgraduate levels. With Iain Biggs, she co-convenes three separate place-based research networks. She has funded various research projects with traditional and indigenous communities, most recently in Brazil with *Caiçara* fishermen, and with contemporary art, as influenced by traditions in China, the USA and the UK, with past research in Aotearoa/New Zealand, Australia and the Isle of Man.

Introduction

In the book *Mutual Accompaniment and the Creation of the Commons*, liberation psychologist Mary Watkins quotes an Aboriginal Activist from Queensland, Australia, who said,

> If you come here to help me, you are wasting your time.
> If you come because your liberation is bound up with mine,
> Then let us work together.[1]

With a sense of kindred intent, we co-authors propose a shared desire to work together in the many ways that are represented here in this book. We aim to move forward as collaborators, colleagues and co-actants, including but going beyond a specialist interest in arts or academic topics. Although our professional experience *is* within the arts and academic research, we hope to encourage you, whatever your background, to understand your skills and knowledge through this book within larger, intra-related ensembles of practices and endeavours. Like Watkins, our desire is to move thinking away from the assumptions of the sovereign self and its hyper-individualism so as to stress "mutual, dialogical, participatory and horizontal relations".[2]

In this respect, we want to contribute to what Anna Tsing, Heather Swanson, Elaine Gan and Nils Bubandt refer to as the *Arts of Living on a Damaged Planet*.[3] But while references to texts by a liberation psychologist and anthropologist help indicate our breadth of concern, this book is ultimately animated by mutual accompaniment and multiple voices, whether alongside students, our peers and colleagues, or through chance encounters with strangers.

It is necessary to be clear who 'we' are. On one level this term refers to those living comfortably in the more affluent areas of the Global North who are inextricably bound into what Amitav Ghosh calls 'The Great Derangement', the consequences of which are politics, cultures and economics bound together in a shifting and rapidly deepening global socio-ecological crisis.[4] One colleague noted that it may be rather like overhauling a ship while it is at seas; and that analogy may not be far wrong. It is also, of course, an authorial 'we', in many instances a duet, different but complementary, as well as a collective voice predicated on those educational opportunities that make a particular form of authorship possible. In our *pas de deux* as co-authors, we recognise the value of our own differences, and wish to convey the usefulness of speaking at times with different language and complementary perspectives.[5] Ultimately, we acknowledge, this is also a privileged position. These limitations we are willing to acknowledge, even if we might wish them to be seen as qualified by the insights we draw from our own individual circumstances.[6]

That said, we would be the first to advocate that readers reflect carefully on the socio-political implications of this text on the basis of your own experience. A form of 'poetic', associative engagement is invited here, offering these words and images as a springboard into your realm of experience and imagination. By implication, then, the plural pronoun might well include you as well. But it never assumes, and is not a spurious slide into a suspension of critical challenge. Your critical reading is essential as the content here demands: in a world with rapidly changing manifestations of the impact of human thinking, the need for careful consideration, philosophical reflection and real action based on this understanding is imperative.

There is, however, a third psycho-social sense in which the word 'we' is used here. This book is the product of a plurality of voices, not simply in the inclusive terms of those who have contributed to it but also with the plurality of voices that flow from our acknowledged natures as complexly constituted individuals and as 'ensemble creatures' in permeable states of being in the world. The 'I' that is a precursor to the 'we' is always changing and adapting, swayed by this and shaped by that. As moments in the social and environmental state of the world shift, so too does this shape the 'we' who are speaking and listening.

Too much writing about the arts and environmental issues today tries to place itself in an unassailable position; to demonstrate its 'rightness' through an exhaustive accumulation of the latest scholarly material across multiple fields. Our ambitions are more modest. As Disabilities Studies within the Environmental Humanities has made very clear, the values people seek to validate (e.g., the value of conservation) can all too easily turn out later to have had, at least partially, a buried and unacknowledged history that deeply intertwines aspects of previous histories with other, deeply problematic, concerns (such as eugenics, in this example of conservation).[7]

Consequently, as authors we claim only to be reporting back as best we can from our own engagement in a multitude of mutual accompaniments we know to be entangled with complications, ambiguities and contradictions.

We co-authors have espoused a kind of 'disciplinary agnosticism' that, while oriented to disciplinary knowledge, should also be extended to this text. As Bruno Latour has made all too clear, our collective future, if we can now speak of such a thing, will require a fundamental rethinking of claims based solely on either specialist expertise or professional authority.[8] Or, to borrow a thought from Donna Haraway, what becomes increasingly important is not claims to authority derived from a particular set of specialist practices or discourses, but from the ability to 'stay with the trouble' in all its many dimensions.[9]

The types of creative engagement with place we discuss in this book are compound, hybrid, involving a multitude of activities, including many different forms of creative practice and research. This distinction is important because, although academically research is usually evaluated on criteria that derive from the sciences, study by anthropologists has shown that the value of creative or arts-led research lies primarily in its ability to generate expertise, confidence, understanding and new orientations to issues, problems, concerns and opportunities – including the production of new conceptual tools and practical abilities. On this basis we would stress that the foundation upon which we write – what authorises us, if you like – is precisely this combination of understanding problems through active encounters and creative practice. The result has been a finding of new creative and conceptual tools that we, and those whose practice is referred to here, share. In this respect, what follows is based on quite a different premise to academic texts, which are underpinned by disciplinary or interdisciplinary scholarship.

To state this comparison in even broader terms, the scientific quest for data that is replicable and verifiable is one which aims to establish truth through proof of a hypothesis. In the arts and humanities, there is a qualitative place for individual experience that stands for the whole: the individual case study offers something useful from which to interpret and analyse experience in the world. Consequently, this latter type of research is concerned with *responsiveness*, seen as an aspect of citizenship in its fullest sense, with opening up: "spaces and opportunities for discussion, argument, critique, reflection"[10] that shape how we act in the world. While this understanding runs contrary to older assumptions about the outcomes of research, it should remind us that attitudes are collectively changing with the same slow but inexorable force as of plate tectonics. This book is itself evidence of yet further shifts in thinking.

The term 'geopoetics' – which we have preferred to the more literary 'ecopoetics' elides 'geologic' with 'poetics', and is central here. It seems to have been coined by the American geologist Harry Hess and subsequently used in a variety of different contexts.[11]

It suggests connections between the earth in its various forms and a poetics of understanding, inherently associative in process, but open-ended, not enclosed in rigidly defined concept or significance: inclusive and unfolding in its meaning and potential. This is an invitation to understand the logic of the earth as poetically construed. Our use of *geopoetics* was suggested by Antony Lyons, an artist who works out of a geo-ecological context, as pointing towards a 'poetics of the earth' and we use it in this open sense rather than to signal any alignment with other disciplinary approaches that use the term.[12]

'Poetics' as an approach is characterized by the ways in which the different aspects and voices of a text and subtext – or in this case, earth as Ur-text – are in inter-relational connection with each other and with the inhabitant/perceiver.[13] Geopoetics, then, is a way of establishing an attentive conversation around all the concerns detailed above, in a manner that allows for scrutiny of the individual aspects of the interconnections between them. We take 'geopoetics' to be an approach to engaging with place in all its aspects from the perspective of 'ecosophy'.[14] In adopting the term ecosophy, which refers to three interwoven dynamic fields or environments: the environment in the usual sense and our place in it, but also the ecologies of society and of self, in the contexts of arts and education, we also have in mind that, in educational terms, the concept of *sophia* – 'wisdom' – is more open and inclusive than the term *logos* – 'logic'. The coinage of words may rest upon the roots of earliest Greek or Latin words, and then invariably have other uses, other people who identify in slightly different contexts or application, and even other precedents that are unrelated to the current intentions. For example, the fragment *phil-* is to love – as in 'philosopher' – literally, a lover of wisdom – and could be joined with *geo-*, as in 'geo-philosophers'. Suffice it to say that we have written here about the love of the earth and all that it represents, and we have embraced the need for deep wisdom. Beyond the coinages of words, which we wish to avoid, and the languages we employ to record these observations, the fullest intention is to embrace with inclusivity and multiple perspectives. Language all too often is unwittingly deployed to be exclusive rather than inclusive; it the aim here to address ideas in language stated as simply as possible with direct clarity and accessibility, even if at times prone to the creative and imaginative use of metaphors, which appeal so broadly to creative thinkers.

The hermeneutics of place is also involved in aspects of this thinking. In this case, interpretation of place is not necessarily a matter of interpreting specific events or actions in themselves, but looking collectively, *in* time and *over* time, collectively and in accruing patterns in nature.[15]

Ecosophical concerns, as they are addressed in this book, imply a process of looking at converging and contested spaces, cognizant of how philosophical, imaginal, theoretical and embodied understandings, the most basic assumptions about the nature of being, inform the ways in which a place is inhabited and the attitudes that sustain the patterns of daily living. Cultural attitudes in everyday life are a part of this investigation, first as it plays out in materials and actions, and second as it betrays underlying beliefs.

Vibrant practices

Among the many theorists that have influenced our thinking recently, have been the New Materialists. We hosted a recent conference on 'Vibrant Matters', and have been collectively discussing Jane Bennett's eponymous book.[16] Added to this are Rosi Braidotti's nomadic theory, Karen Barad's work on agential realism, Manuel De Landa's writings on material culture, and Iris Van der Tuin's ontologies of new materialism. This has opened out an analogous attention to what we might call 'vibrant practices' in art, design, landscape architecture and environmental interventions. This body of theoretical concerns and more prompts an approach to the host of practices referred to based on a 'joined-up' thinking that understands the environmental, social and personal as part of a constellation of connectivity that is ultimately woven, however loosely at times, both dynamic and 'all of a piece', based in time-space-matter. In the chapters that follow, many of these individual creative practices are traced through interwoven concerns of vibrancy, connectedness and shifting interrelations in flux.

Forms of these practices vary with each project, individual and collaborative. One method of documenting these complex activities, particularly as they are connected to sites, is through 'deep'-, 'narrative'-, 'fluid'- or 'slow'-mapping, a compound and shifting term discussed at length in Chapter 1. Experiments in such mappings are referenced throughout this book, with examples in several chapters, but it is important to stress that there is no one approach to the representation and collection of multiple aspects of understanding that overlay and inter-combine. Every conventional 'map' begins with basic assumptions and the intention to represent selected aspects of knowledge. These alternative modes of mapping challenge some of these assumptions, and seek to counteract the inaccuracies of looking at information that is too narrowly defined or depicted. Inter-relationships, and dynamic interactions, are the focus of this emergent technique (or more properly, emergent range of evocations, visualisations and enactments).

It is with this commitment, then, to plurality as a way of understanding being in this world, to examining dynamic shifting interconnections, and to the representation of a profound respect for the earth itself, for animate and inanimate fellow 'creature-objects' with whom we share this planet, that the authors have engaged in this undertaking.

Outline of contents

This book begins in Chapter 1 with an overview of the territory: the nexus of art, education and place (plus place-based research and theory) as the ground for new thoughts about ecosystems of thinking and acting. Historical actions, theoretical writers, indigenous belief systems and Western philosophy are briefly reviewed to lay a foundation for this work. It then explores two core terms: 'geopoetics' and 'deep' or 'slow' mapping, offering these as new, open-ended tools for thinking that the reader can then adapt to her or his own thinking and practice.

Chapter 2 considers how we take in the world – how perception and interpretation lead us to the simple most basic actions of Seeing, Listening and Acting. Along the way, an imaginary walk with the reader looks at the visible and the invisible, the phenomenal and the numinous. The chapter asks the reader to reconsider that which is readily available to sensory modes of perception, and not to dismiss other, less obvious ways of knowing in understanding the importance of sited knowledge.

Chapter 3 returns to issues of deep-mapping, and looks at many creative practices that engage with the environment and place-based research through 'slow residencies' and sustained commitments to long-term efforts and investigations.

Chapter 4 looks at the lifelong practice of artist Christine Baeumler, examining her 'Call to Eco-Action' with citizen-scientists, community groups and children through a process of collaboration and education through activities.

Chapter 5 investigates the values inherent in perception, and especially as this is central to the ways in which place is perceived. Threads of cultural values, which are attendant to one's understanding of a given site, are pulled, strand by strand, from the complexly constructed fabric of understanding.

Chapters 6 to 9 set out some of the practices and core concerns of people who are active in creating vibrant practices and slow residencies, working in the environment with aims to address various aspects of ecosophical concerns. We address creative communities of practice in Chapter 6, and emergent practices from new interdisciplinary geo- artist-activists are featured in Chapter 7, Gini Lee's Curation of the Land in Chapter 8 and Alexander and Susan Maris are addressed in Chapter 9. All address communities of transverse action through their work, and in substantially different ways.

Chapter 10, which deals with the central issues of 'fieldwork' and *notitia* – understood as a particular form of attention or listening – serves as a form of 'anti- conclusion', aiming to return the reader to exploring her or his place in the world, and the world's place in shaping them. The promotion of active creativity, of applied intelligence, and openness to the shifting demands of a changing environment are interwoven here. And in place of a summary, the final pages propose a move towards ecosophical collaboration and an expanded frame of reference for practitioners of many types.

Intentions and action

What we present in this book as authors reflects on (and seeks to encourage the reader to consider) the relationships that inform ecosophical practices as open-ended; rough rule-of-thumb approaches that she or he can adapt in the light of circumstance and need. In this respect it might be compared to a book like Mary Watkins and Helene Shulman's *Towards Psychologies of Liberation*, with which we share the view that: "to create cultural alternatives, people have to break with taken for granted ways of thinking that prevent them psychologically from interrupting the status quo".[17]

While this is in not a 'how to do . . .' book in any conventional sense, it *is* intended as an aide to practical action in relation to questions of place. It is a prompt, an appeal for creative engagements of your own. What is offered are orientations, attitudes and intentions to act, as much as accounts of forms of creativity off to one side of the mainstream. Our hope is that they will serve as an inspiration that you, reader, can pick up and take further in your own place. The examples and ideas offered here are dependent on a mode of attention that employs a site-specific critical poetics to address the particular web of inter-connections that makes up any particular portion of the life-world as *taskscape*.[18] And for as much as we avoid neologisms, there is a merging in this understanding of place and things needing attention and concerted action.

They do so by interweaving evocations of its material, temporal, spatial, imaginative and conceptual dimensions in a largely non-hierarchical manner (with the exception that ethical choices always require a degree of hierarchical judgment at the level of value), and in terms of an ecosophical understanding. It follows that they are never 'ours' as in the ownership of a brand but are always: 'made up of – and physically manifesting – other peoples' work, input, substance and knowledge' as well as our own. This in turn means that: "there is no project that is not already the project of other people as well".[19]

In consequence you, dear reader, are already implicated in this project by virtue of having followed our thoughts this far. And we invite you to proceed in this journey.

Notes

1 Mary Watkins, *Mutual Accompaniment and the Creation of the Commons* (New Haven, CT and London: Yale University Press, 2019).
2 Watkins, 2019, 1.
3 Anna Tsing, Heather Swanson, Elaine Gan and Nils Bubandt, *Arts of Living on a Damaged Planet* (Minneapolis, MN: University of Minnesota Press, 2017).
4 Amitav Ghosh, *The Great Derangement* (New York: Penguin, 2016).
5 And like all dancers, step on each other's toes only occasionally.
6 By awareness flowing from partial Native American heritage in Modeen's case and by Biggs' 25 years of experience as a career for a chronically sick daughter.
7 See, for example, Sarah Jaquetta Ray, 'Risking Bodies in the Wild: the "Corporate Unconscious" of American Adventure Culture', in *Disability Studies and the Environmental Humanities: Toward an Eco-Crip Theory*, ed. by Sarah Jaquetta Ray and Jay Sibara (Lincoln & London: University of Nebraska Press, 2017).
8 Bruno Latour, *Down to Earth: Politics in the New Climatic Regime* (Cambridge, MA: Polity Press, 2018), 15–16.
9 Donna Haraway, *Staying with the Trouble: Making Kin in the Chthulucene* (Durham, NC: Duke University Press, 2016).
10 James Leach and Lee Wilson, 'Enabling Innovation: Creative Investment in and Arts and Humanities Research', in *Creativity and Cultural Improvisation*, ed. by Elizabeth Hallam and Tim Ingold (Oxford & New York: Berg, 2007), 99–118.
11 Harry Hess (May 24, 1906–August 25, 1969) is considered a revolutionary figure in earth sciences and a 'founding father' of the unifying theory of plate tectonics. When he first published his theories and findings in the article 'History of Ocean Basins', he called it "an essay in geopoetry". Harry Hammond Hess, 'History

of Ocean Basins', *Petrologic Studies a Volume in Honor of A. F. Buddington* (Princeton, NJ: Princeton University, November 1962), 599–620.

12   We also draw separate distinctions from Kenneth White's Scottish Centre for Geopolitics, although of course we share several overlapping concerns and interests.

13   A phrase generally understood as referring to 'original' (most ancient), as in the first and most authoritative version of a document.

14   'Ecosophy', as first defined by Norwegian Arne Naess, was described as: "a philosophy of ecological harmony or equilibrium. A philosophy as a kind of *sofia* (or wisdom), is openly normative, it contains both norms, rules, postulates, value priority announcements and hypotheses concerning the state of affairs in our universe". Quoted in Alan Drengson and Yuichi Inoue (eds.), *The Deep Ecology Movement: An Introductory Anthology* (Berkeley, CA: North Atlantic Publishers, 1995), 8. It is often used as synonymous with 'ecological wisdom'. The term was synchronistically used by Guattari in a slightly different manner when he "postulate(d) the necessity of founding an 'ecosophy' that would link environmental ecology to social ecology and to mental ecology". Felix Guattari, 'Pour une refondation des pratiques sociales', *Le Monde Diplomatique* (October 1992), 26–27.

15   This importance of time in the ways of understanding, as well as the frames of understanding, is an issue to which we will return throughout this book. In many ways, place and time are inextricable.

16   Jane Bennett, *Vibrant Matter: A Political Ecology of Things* (Durham, NC: Duke University Press, 2010). The conference took place on 29th and 30th January 2015 at the University of Dundee, Scotland.

17   Mary Watkins and Helene Shulman, *Towards Psychologies of Liberation* (London: Palgrave Macmillan, 2008), 8.

18   'Taskscape' is a term that is credited to Tim Ingold, a social anthropologist based at the University of Aberdeen. He located the term within place-based activities: "just as the landscape is an array of related features, so – by analogy – the taskscape is an array of related activities". Tim Ingold, *The Perception of the Environment: Essays on Livelihood, Dwelling and Skill* (London: Routledge, 2000), 95.

19   James Leach, 'Creativity, Subjectivity and the Dynamic of Possessive Individualism', in *Creativity and Cultural Improvisation*, ed. by Elizabeth Hallam and Tim Ingold (Oxford & New York: Berg, 2007), 112.

1 Geopoetics in context

In the Introduction, we signalled our concern to develop the ensemble understanding we see as the 'fourth ecology'. This term references Felix Guattari's essay *The Three Ecologies*, in which he proposes a new way of thinking, an 'ecosophy' as he refers to it; one that avoids the philosophically rigid way of thinking that strict disciplinarity promotes, by engaging simultaneously with the three interrelated ecological fields of the environment, society and the compound self.[1] In certain respects, our concerns here run parallel with Guattari's notion of ecosophy but, for reasons explained below, we will draw only sparingly on his thinking, preferring to identify our concerns with the field of a geopoetics seen through the lens of mutual accompaniment rather than one that continues, however unwittingly, to replicate the presuppositions of possessive individualism.

As individual authors, we have long experience both of making art in that variety of contexts rather misleadingly called the 'art world'. (There are, of course, many different 'art worlds': commercial, popular, museums and otherwise.) Equally importantly, we have been involved in teaching and research as part of the very different world of universities, the 'academic world'. Living in this way 'between worlds' so to speak, we have grown increasingly concerned about how terms like 'art', 'education' and 'place' are still assumed to conform to an increasingly archaic and irrelevant *disciplinary* understanding, one ultimately grounded in an unhelpful assumption of division and exclusion. Our sense of the problems raised by disciplinary understanding reminds us of the feminist writer Audre Lorde, who argued that using the concepts created by a system to examine the products of that same system prevents real change. The implication is that if the presuppositions underpinning concepts, taken as the tools with which we think, don't change, then creating new terms without challenging the old presuppositions is unlikely to change anything either. This is for us the sense of Lorde's famous statement "*the master's tools will never dismantle the master's house*" (the title of her address at the 'Second Sex Conference' in New York delivered in 1979). Today, the way in which terms like 'art', 'education', 'resilience' and 'ecology' are used too often remains at root in conformity with the values and assumptions of a global economy that, in turn, is underwritten by the presuppositions of the dominant culture of possessive individualism. It is this culture, and the economics inseparable from it that, alongside the culture of a pervasive patriarchy which Lorde calls 'the master's house', currently provide the foundations upon which the dominant culture has been built.

This book starts, then, from a simple presupposition. The dominant social order ('the master's house') that has been built on possessive individualism has become so problematic, so toxic, that it is destroying not only the fabric of human society but the ecologies upon which all living things depend. This is closely linked to the way disciplinary thinking analyses the world by dividing it up and sorting the resulting divisions into categories that stress difference rather than inter-connectedness, and fixed definitions rather than fluid and changing understandings. This process of fragmentation, which has immense personal, social and environmental implications, in turn makes it too easy for individuals to disregard the consequences of their actions for other human and non-human beings. If we are to think and act in ways that address this lack of connectedness, we have an urgent obligation to find a way, if not to dismantle 'the master's house', then to radically remodel it so that we can live quite differently in the world. However, it is important to acknowledge that there is no ambiguity in Audre Lorde's image. For her the tools that belong exclusively to the master are only fit for his purposes. The clear implication of her stark image is that we will need to create an entirely new set of tools if we want to create real change. That would mean entirely new forms of art and an entirely new model of education. Hers is a powerful image, and there is some truth in what she says, but it doesn't necessarily reflect how such cultural tools are used in practice. Tools, in this case educational and artistic concepts and processes, are certainly often used in ways that sustain the power and authority of the *status quo*. But they can also be adapted, recombined and even radically reconfigured and then re-used for other purposes. Two examples of such changes would be Paulo Freire's educational work that informed his book *Pedagogy of the Oppressed* and Joseph Beuys' interweaving of art, education and environmental politics.[2] So, while Lorde's statement forcibly reminds us that we may need to question or even abandon concepts and processes currently used to reinforce existing power structures, we may also need to transform them into new tools with which to think and work. This requires remaining open to alternative, often fuller, understandings which reject an overly literal or binary usage, allowing a translation to serve other ends. Only in this way can we grasp the full range of nuanced associations that link ideas and practices back to the paradoxes, and so to the untapped potential, of lived experience. Working with a combination of vigilance and openness, an approach we call 'disciplinary agnosticism', is central to what we are proposing here.

One aim of this book is to suggest to the reader some of the tools we believe are necessary to do this, which we see in terms of a new, open-ended geopoetics that we equate with Bruno Latour's conception of a Terrestrial politics. At its core, a responsibility of individual accountability is established. It can be argued that an even fuller assumption of responsibilities will be built upon the recognition that the contradictions and shifting relations between the three ecologies of culture, the environment and the individual are often successfully articulated by works of art which, like a good teacher, encourage us to recognise the fluctuations and tensions between them. Of course, the practices of art and education are also economic and political activities, with all the complexities that can flow from that, but this does not negate their importance as means to evoke understandings that oppose or resist the values of the dominant, economically driven, culture. It is important, then, that we avoid adopting dogmatic or factional positions regarding the value of different creative approaches. This is something artists often find difficult to do. If we're involved in 'socially-engaged art practice', for example, we need to remember that painters may have interesting and relevant things to show us, just as we need to remind ourselves that not all socially engaged projects are interesting or relevant. Above all, we need to keep in mind that tension and paradox are as central to the creative arts as they are to our lives.

That being the case, we need to identify a paradox central to this book. On one hand, both art and education are manifestations of what is often claimed for them; namely, that they are vehicles of independent thought, discrete forms of cultural activity and semi-discrete domains of activity that contest given values. But they are also what they often pretend not to be: "bound up with values of the status quo and the ideological system that sustains it".[3]

Consequently learning, whether in an institution, by reading a book, or engaging with art, can be *both* a process by which we are further conditioned into the knowledge, patterns of thought and values privileged by the dominant social order *and* an opportunity to question the presuppositions that underpin that knowledge, those patterns and values. We find this difficult to accept because the heritage of the three monotheistic Religions of the Book (Judaism, Christianity and Islam) leads us to believe that we act productively by making judgments based on predetermined and fixed notions of what is 'right' and 'wrong'. But as compound or ensemble selves, we need to learn to live and work with paradoxes and contradictions in a world that is a lot less clear-cut than religious or secular dogma and ideology suggest.

A book like this inevitably raises questions about the links between language and exclusion, as we suggested briefly in the Introduction. If we want to engage with as broad a range of readers as possible, we need to use accessible language, not least to combat the use of hyper-specialist professional, technocratic, or disciplinary terminologies as exclusionary. But, equally, we need to draw on, and put in play, new ideas (and new tools) if we want to facilitate alternative possible ways of living and working. We have tried to use as language as clearly we can, without shying away from the new ideas we find necessary to go beyond the disciplinary *status quo*. We recognise that this approach risks displeasing everybody, with professional artists and academic specialists seeing us as simple-minded popularists, and general readers as obscurantist. It is a risk we are willing to take in adopting a new approach. We can only hope that readers will be sympathetic.

The issue of language comes up as soon as we think about place, often taken as synonymous with what is static, given and bounded. Edward S. Casey points to an issue of fundamental importance when he reminds us that:

> if a position is a fixed posit of an established culture, a place, despite its frequently settled appearance is an essay in experimental living within a changing culture.[4]

The question of ecologies of place central to this book is related both to Casey's thinking and to practical changes in, for example, the architectural practices of Critical Regionalism and its subsequent developments.[5] This shift to sensing ourselves as always placed dynamically (rather than merely positioned) in the world, also requires us to question what it means to 'know our place' in terms of the various ecologies that make up our worlds. Although the political geographer Doreen Massey is dismissive of Casey, largely due to his use of the philosopher Heidegger's thinking, much of what she writes about *space* chimes with our understanding of *place*. For example, its being produced by interrelations and interactions that generate: "the possibility of the existence of multiplicity in the sense of contemporaneous plurality", and being best imagined as "a simultaneity of stories-so-far".[6]

Thinking about 'our' place in these interrelated senses requires that we question two dominant current presuppositions. First, we need to question the supposition that what makes an individual unique is her or his difference and separateness from others. Second, we need to question the idea that human society is something distinct and separate from the 'natural' world. Ecosystems, whether at the level of micro-habitats or bio-regions, are (almost always) open; that is, sustained by a continual flow or flux of energy and matter across the semi-permeable borders between them. If places seen ecologically are best understood as consisting of "a polyrhythmic composition of processes whose pulse varies from the erratic flutter of leaves to the measured drift and clash of tectonic plates" and as "a tangle of interlaced trails, continually raveling here and unraveling there", then to think and act ecosophically requires a similarly open, flexible, multi-scaled and fluid approach to both our sense of self and the relationship between the human and the non-human.[7]

As inspired by the actual boulders in the Pentland Firth, in the *Nomadic Boulders* by Dalziel and Scullion (Figure 1.1), the 'place' of these stones is a nexus of time-site movement, subject to the impact of their environment with powerful motions of sea currents. In relating this lack of fixity back

Figure 1.1 Dalziel and Scullion, *Nomadic Boulders*, (2015). Photo: the artists. This work is sited in John O'Groats, where some of the fastest sea currents in the world have been recorded and in which a number of large rolling boulders are known to traverse back and forward on the seabed.

to most contemporary education (including art education), unfortunately there is still a presupposition of a disciplinary approach grounded in an exclusive mode of investigation that works against thinking *across* time-and-place. It is for this reason that Guattari's counter-disciplinary notion of relationality, of an ecosophy of three interrelated ecologies of environment, society and self, is so relevant.

This book enacts the thinking necessary to the understanding we refer to as the 'fourth ecology', a term also discussed in detail below. Our approach has been to build on our own experiences as a loose collective of compound selves long engaged with, among many other activities, the arts, their theory and philosophy, education and ecology, all in various ways and forms.

For reasons that should be clear by now, this is neither an academic text-book nor an 'art book' as these are usually understood. Instead, it com-bines some of the functions and conventions of both. In addition to being an informal manifesto of sorts, it is intended to be a point of reference for a range of ideas, practices, histories and contexts, including an argument for the importance of 'fieldwork', all ultimately seen as necessary components of any genuine democracy.[8] The result is a loose weaving-together of related and mutually informing strands of material, rather than a linear argument designed to deliver a particular conclusion or an authoritative overview.

Disciplinary agnosticism

In this book, we have situated ourselves off to one side of disciplinary think-ing, that is: "the general disciplinary project of producing and regulating the movement of knowledge, the forms of language, and the training of minds and bodies".[9] Our alternative position is one of 'disciplinary agnosticism', which is, put simply, a deliberate suspension of 'belief in' the authoritative claims that the whole system of disciplinarity assumes. This does not, of course, mean that we dismiss the necessary and valuable skills, crafts and understandings we learn from discipline-based study. It means, rather, that we need to see disciplinary knowledge and practice as simply one among a whole range of other useful and constantly mutating processes and ways of understanding. What we do reject are certain characteristics of disci-plinary culture; its manipulative fragmentation of knowledge, its drive to exclusivity, and its emphasis on hyper-specialisation. These are attributes re-enforced by social and institutional pressures linked to possessive indi-vidualism: to over-invest in a professional identity – that is, in 'my' scien-tific discipline, 'my' exclusive art practice and so on, at the expense of all other qualities. When this over-investment becomes a secular belief system, the techno-optimism of eco-modernists for example, it stops us thinking productively and becomes a form of socially constructed blindness or igno-rance, the basis of a 'life-as' in opposition to a 'life in. . .'[10]

Our disciplinary agnosticism flows in part from working with people with a depth of knowledge, experience and a different cultural perspective *that sits outside disciplinary thinking*, work that has changed our own un-derstanding and practice. This experience of an unsettling of the taken-for-granted disciplinary framing of western education and cultural knowledge through a profound encounter with another, non-disciplinary world-view is something that we share. It provides one footing for our relationship to the larger cultural shift already mentioned in relation to place and reflects a more general welcoming of the thinking and practices of non-professional or non-disciplinary constituencies. These are the groups that tend to be largely invisible in an academic world that tacitly presupposes all worth-while knowledge as discursively framed and located within a disciplinary hierarchy. To adopt a position of disciplinary agnosticism is, then, to sus-pend belief in (but not dismiss out of hand) the assumption of authority made by, or on behalf of, any one intellectual or professional discipline.

We assume here that it is *as important* to understand the connectivities and discontinuities *between* and *outside* disciplines as it is to understand what any individual discipline tells us. This approach helps illuminate 'how the world works' in practice, with a view to changing it, and counters the professional tendency to over-invest in any one exclusive mode of producing and manag-ing knowledge. This shift is vital to moving towards an expanded, ecosophi-cal thinking, helping us to remember that a discipline is not a given or natural phenomenon, but a model that is both historically and socially contingent.

In this context, it is important to add that disciplinary agnosticism is understood here as an intellectual, not an ethical, position. It proposes an agnosticism to the professional claims made by academics, artists, environmental scientists and others whose authority derives from discursive, disciplinary, knowledge. It does not excuse us from making and acting on ethical judgments about the balance of rights and wrongs in any situation. Our adoption of disciplinary agnosticism means that, unlike many of the authors we have learned from, we will not be arguing for a new, expanded form of disciplinary practice. Instead we want to promote an agnosticism towards all claims based on professional or disciplinary discourse; one that, in the spirit of Ivan Illich's *Disabling Professions*, we see as vital to keeping in question the *realpolitik* of professional power and authority.[11]

Our adoption of disciplinary agnosticism is also in part prompted by the feminist philosopher Geraldine Finn's observation that: "we are always both more and less than the categories that name and divide us".[12] We take this to mean that we cannot simply assume the authority of artist/academics when we write but, instead, must reflect the larger, messier, more complex and multifaceted nature of our lived experience, taken as that of compound or ensemble selves. That is, we must write as people engaged in the open-ended project of becoming better Terrestrial citizens, while simultaneously remembering that we are each always embedded in the connectivities and disjunctures of specific places, times, histories and the practices and contingencies central to them. As fallible, contradictory and imperfect beings, we understand the need not to claim some exclusive or 'new' authority here. Readers will need to assess the value (or otherwise) of the material offered in this text in the light of their own experience and understanding.

Disciplinary agnosticism in practice

What does disciplinary agnosticism mean in practice? The American art historian Miwon Kwon is best known for her book *One Place after Another: Site Specific Art and Locational Identity*. That book serves here as an indicative example of the discourse of a whole category of professional persons, one that includes both artists and academics. That category is, in turn, part of an international elite able, in Kwon's words, to measure the viability and success of their work 'by the accumulation of frequent flyer miles' in the process of providing national and international institutions with their 'presence and services'.[13]

In 2012, when reflecting on the claims of Kwon's book, I read about Consolata Melis, a 106-year-old Sardinian woman who, with her eight siblings and 180 living descendants, made up a substantial percentage of the population of a remote town, concerned largely with the business of sheep-farming, in the mountainous region of Ogliasta. Consolata Melis reminded me of all the people who are rendered invisible by the type of discursive theorising practiced by professional and academic elites; of all those *others* whose lives do not necessarily conform to statements such as: "*Everybody* [author's italics] speaks to, and answers to, an invisible ear, one that belongs to a phantom body of a televisual public".[14] Those *others* whose lives put in question the assumption that it is: "historically inevitable that *we* [italics mine] will leave behind the nostalgic notion of a site and identity as essentially bound to the physical actualities of a place".[15] Exercising disciplinary agnosticism here reminds us that 'we' (the critical plural) should not assume the right to make such exclusive claims, particularly when they relate to worlds, like that of Consolata Melis, that function on the basis of presuppositions that are now almost entirely unfamiliar to the dominant culture.

Consolata Melis can also serve to remind us that the sense of communal and individual identity of a substantive percentage of the world's agricultural producers, including a not-insignificant number in the United States of America and Europe, is necessarily bound up with the complex physical actualities of a place as a lived mesh of relationships with both human and non-human entities. As a friend said recently about her Slovakian daughter-in-law's family: "They all have plots of land on which they grow all their own produce, keep chickens, rear bulls. Even the granny, who is eighty-one, digs her own potatoes".[16] It should not be necessary to point out that it is precisely this intimate knowledge of the physical actualities of place, as these relate, for example, to managing water, animal husbandry and growing crops (among other skills), which provides many people with a living, one through which they also contribute to the agricultural production upon which we are all ultimately dependent for our daily food. In a recent conversation with a Brazilian fisherman who has two young children, he stated, "When I teach my son to fish, he will never need to buy a fish in the market".[17] The rural world of food production may be all but invisible to those who measure their success in terms of the frequent flyer miles, and to their assumed audience of largely urban-based professionals, but it is none the less real for all of us who eat.

In the example above, disciplinary agnosticism regarding the claims of an art historian and theorist helped me keep an open mind regarding the assumption of authority that underpins her theorising about the relationship between place and identity. My observations are not, of course, intended simply to dismiss Kwon's argument, which is, given its tacit assumptions and terms of reference, persuasive enough. Rather, they are intended to reflect agnosticism regarding the assumptions upon which she stakes her claim to speak authoritatively for an 'us' (the 'we' of her text). By adopting that agnosticism, we can better 'place' her argument; we see it as informed by her identification with curation as a profession, as a resident of urban California, and as someone who has internalised certain values dominant in the cultural *milieu* in which she lives and works. It is this *milieu* that also provides her with those self-selected peers – artists, art historians, theorists, curators, other art professionals and students who, along with members of the 'art public' – that constitute the assumed 'we' of her text. It is a simultaneously inclusive pronoun for her culturally similar sub-group and, as such, tacitly exclusive of most others. The lived experience of rural worlds neither conforms to, nor invalidates, Miwon Kwon's argument. It simply reminds us that we live in a polyverse.[18] Disciplinary agnosticism is not primarily concerned with 'taking sides' or 'winning arguments', although these may be necessary secondary concerns, but rather with better *placing* and *understanding* the conditions that frame our assumptions.

A community of transverse action

The reader will have noticed that, in the previous section, I used the first person singular rather than our usual authorial 'we'. I did so because I was writing from my own experience, but also to raise with the reader certain questions about authorship. That most books have a single named author should not blind us to the fact that they are in actuality the product of work by a multitude of people, not least all those upon whose knowledge and experience an author builds. This book is best described as co-authored by us two, Modeen and Biggs, but broadly speaking as the product of a community of transverse action, one whose members and contributors earn their living to a greater or lesser extent through multiple practices in the arts, architecture, landscape design, academic research and teaching, and whose concerns are inflected by various forms of social engagement and activism. As a community we share overlapping interests – particularly in creative and ecological issues – but are geographically located across three continents. We refer to ourselves as a 'community of transverse action' to indicate that we are a self-configuring and fluctuating community that appears because of, and is united by, a commitment to learning-as-becoming, a process enacted through mutual support, enquiry and discussion oriented by an attitude of critical solicitude. Each member of this community has her or his own aims and concerns. What we share, however, is a desire to make socially productive ecosophical interventions in the world on a variety of scales. Both our individual and collective activities are undertaken through an overlapping set of skills and concerns, locating us somewhere between what is conventionally understood by a 'community of practice' and a 'community of interest', as these terms might be reinterpreted in the light of Felix Guattari's thinking.

This transverse community is distinct from the informal disciplinary or interdisciplinary academic and professional networks it somewhat resembles (and with which it inevitably overlaps). Its stress is on lateral thinking and acting ('transversality') and on openness to non-disciplinary understandings. As a loosely interwoven transverse-community we share a commitment to 'transverse action', seeking to bypass the restrictions imposed by institutional norms and hierarchies, to engaging practically with non-professional, vernacular or 'lay' knowledges and, above all, to learning-as-becoming undertaken as a collective, open-ended endeavour. This filters out into other constituencies and communities through larger networks of people who value arts and design skills and processes, including those of deep mapping/slow residency, as social practices. It seems, then, that many like-minded people have begun to recognise that some version of this process of mutual, conversational, outward-facing, self-education is central to their lives. The interlinked processes central to such a community parallel, in many respects, the mycelial mesh of relationships between material environments, broad social relationships and the inter-subjectivities of human (and on occasion non-human) beings that make up our understanding of place.

The heart of this book

Our aim as authors is to evoke a 'fourth ecology', a dynamic intermingling of relational knowledges and understandings, which, as a community of transverse action, we see as flowing from our concerns with, and practice of, geopoetics, deep mapping/slow residency and the ecologies of place. This aim is animated by what the writer Alan Garner refers to as the education system's privileging of analytical disciplinary thought above all else, preventing it from fully engaging with cultural practices as a legitimate form of understanding, a system that cannot acknowledge the paradox of "the integration of the non-rational and logical" that "engages both emotion and intellect without committing outrage to either".[19] As already indicated, our collective experience requires us to embrace this form of paradoxical understanding, found in a poetics grounded in ethical, philosophical and ecological presuppositions about the place of all living creatures, human and non-human, and the conservation of the earth's finite resources.

Ecosophical thinking allows us to see that 'nature-society relations' are not just a one-way process; to acknowledge that: "'nature pushes back' with its own vitality which is manifest in specific material processes".[20] This situation means that we need to attend to the agency of the earth itself if we wish to survive and, because there is an important connection between "understandings of agency and understandings of ethics", we can only address our personal and socio-ecological situations if we develop and put into creative practice these alternative ways of knowing and understanding the world.[21] This is also what it means to address the issue of possessive individualism. It is significant that these alternative understandings share in some degree a way of thinking about reality that is: "not far removed from understandings of the lifeworld professed by peoples commonly characterised in ethnographic literature as animists".[22]

This is significant because exchange with indigenous people in North America, Australia and New Zealand has been important to us. While we cannot simply unlearn our current knowledge as beneficiaries of the Great Derangement, we can learn a great deal from what our culture has either dismissed or sentimentalised as 'archaic' or 'animistic' traditions. Moreover, we are aware of the practical efficacy of many indigenous traditions that have been better able than contemporary industrialised society in achieving a more harmonious balance with their environment. Knowledge of these traditions enables us to reflect on the dominant mind-set by making concrete the consequences of the fact that: "ecological disaster, which we know about only because of very sophisticated interdisciplinary science, has torn a giant hole in the fabric of our understanding".[23] This realisation is generating an ecosophy that envisages interconnectedness or a meshwork in which nothing and nobody 'exists all by itself', so that 'nothing is fully itself', but is always porous, semi-permeable to what is normally viewed as 'other'. It is this understanding that sharply distinguishes what might loosely be called our ecosophical, neo-animist thinking from the culture of possessive individualism.[24] As we have made clear, to be able to put such thinking into practice requires other ways of knowing and creating. This brings us to our observation that signals both the risk involved in our approach seen from an academic viewpoint, and the necessity of taking that risk:

> To those of us who are academics . . . love is a four-letter word. It is immeasurable and therefore by its very nature outside academic territory. It cannot be calculated, predicted or even adequately defined except perhaps normatively, as enacted by individuals, new in its manifestations each and every time. Even though it is as old as humans themselves, probably predating that which we know as human, shared (we are certain) by fellow creatures in the animal kingdom, and known as well as 'the force that through the green fuse drives the flower', it is not *academic* properly speaking, not to be trusted, best avoided for other less risky terms. And yet . . . it is the best word I can think of to discuss the ways in which we interact with our environment.[25]

What is undertaken here is 'done for love', requiring us to simultaneously and paradoxically address *and* circumnavigate the difficulties of 'writing about' and 'showing' new modes of 'knowledge production' through cultural work. This should not be confused with what, in academic contexts, are referred to as 'post- disciplinary' or 'relational' approaches. What we are centrally concerned with here is the power of imagination, with what was formally known as: "*himma*, the thought of the heart" (Figure 1.2).[26]

Working with paradox in practice: *both* arts *and* the sciences

Following Alan Garner, we might say that this book is based on the paradox already indicated. It presents, contextualises and reflects on examples of a loose weave of creative practices that deploy a critical poetics to engage with the three ecologies of the spatio-temporal landscape we inhabit (our 'place'). These practices draw on both the creative skills and methods associated with the arts; those, for example, of the photographer, printmaker, painter,

Figure 1.2 Dr Amy Todman, *Speaking to Lichen*, photograph of performative actions, Scotland, 2015. As a performative artist, her (art) actions typify
the listening and speaking inherent in true conversation, and also as a
loving action. Photo: Amy Todman.

film-maker, landscape architect, sculptor, performance artist and writer, but
also on a wide range of disciplines and practices within the humanities and
the social, earth and environmental sciences. Drawing on Kenneth White's
notion of geopoetics and Mike Pearson and Michael Shanks' argument for
going beyond the art/science polarity (in their *Theatre/Archaeology*), this
dispersed community of transverse practices presupposes that the categor-
ical differences between the arts and sciences are, at the very least, permeable, and at its, profoundly overlapping and interconnected. Both, for
example, are at best creative practices requiring a combination of observation, imagination and analytical discipline.

The purpose of Chapter 3: Deep mapping/slow residency, is to give the
reader some sense of how one cluster of practices, broadly those within traditions of deep mapping/slow residency, may be seen in terms of their role
within a larger geopoetics, However, while not all histories are written 'by
the winners about the losers', there is always a danger that a strictly historical account will place too much emphasis on canonisation, on 'who did what
first' and what is within, and excluded from, consideration. Rather than a
history, or even historicity, of deep mapping/slow residency, Chapter 3 aims

instead to provide the reader with a sense of the variety of possible models and resources available, based on a sense of 'familial' relations. First, however, it is necessary to indicate how these models and resources are framed by our own concerns and values. The remainder of this chapter sets out to do this.

We have adopted a deliberate blurring of disciplinary categories by, for example, conflating and cross-referencing literary and visual arts genres, social science disciplines and more. Our justification for this is again neatly articulated by Alan Garner, who writes that: "It may be that the purely academic mind" – a mind which insists on the primacy of analytical categories over intuited affinities – "will always be wary of the eclectic, deeply ordered chaos of the maker", since the artist "will always and instinctively resist the scholar's quest for the finite answer".[27] In a culture in which the dominant mind is academic and the analytic mode concerned with tools and data, methodology as order and control are primary; it is both proper and necessary to promote a counter-balancing thinking based on the broadest intuited and practical affinities.

Geopoetics, deep mapping/slow residency: locating relations in place/time

The historical and geopolitical contexts that frame the currently dominant form of thinking in 'the West', along with activities such as cartography that flow from it, cannot be wholly disaggregated. Because both our thinking and practices have "contingent roots in particular persons, places, and times", when we think of making maps we tend to think only in terms of a cartographic tradition rooted in the needs of predatory merchant traders, the military and colonial administrators.[28] Even if we are aware of the maps produced in other traditions, we tend to think of these as 'not proper maps'.

To most people, maps in the tradition of the Indian sub-continent that informs Gulammohammed Sheikh's installation: *Kaavad: Travelling Shrine: Home* are exotic curiosities that have no practical relevance today. However, to look thoughtfully at such maps is precisely to get a sense of the limitations of mainstream Euro-American cartography. In their own terms, these are images of great explanatory power to those able to read them in their proper context, which is as much spiritual and social as it is cartographic in our limited sense.[29] And, if we so wish, we can find examples of a similar, if less expansive, metaphysical response to the cartographic urge in the recent past of our own culture. We stress this because, in the contemporary West, it is now conventional to see the aesthetic and the cartographic image as distinct from, even as opposed to, each other, with the former being a 'subjective' and the later an 'objective' representation. However, this polarisation merely reflects Western culture's epistemological presuppositions.[30] For Aboriginal Australian people such as the Warlpiri, maps of their Dreaming are, among other things, facets of a cultural ecology around which their society is organised; in short: 'paintings are maps of land'.[31] But equally, because their Dreaming is situated, is conceptual and affective as well as spatial, must be read as complex codes for multidimensional experience. Therefore, we have to avoid any wholesale transfer of Western topographical concepts onto such works if we want to understand them properly. In actuality such works make complex conceptual representations of place and space that respond both to mythology and to social relations that may involve 'exercising rights of inheritance and ritual authority' as well as implying 'rights in the land itself', of necessity needs a substantive psychosocial, legal and ritual knowledge.[32] Arguably such works simply do not conform to our division between 'subjective' and 'objective' ways of knowing. This is the position that is replicated in deep mapping/slow residency. Openness to the potential of different cartographic traditions across a variety of cultures, seen as a resource for visualising complex relations, is clearly one real benefit of acknowledging the contingency of our own presuppositions.

While it might be argued that a substantive number of non-Western and anti-mapping traditions feed into deep mapping, there are obvious limits to how far these numerous possible roots can be identified in a book of this kind, let alone followed back across place and time. As Karen Till rightly observes, even in contemporary academic circles there are many varieties of alternative mappings, all forming part of an ongoing project that, in addition to those activities mentioned in the Introduction, include: 'memorial cartographies', 'haunted archaeologies', 'experimental geography' and 'radical cartographies'.[33] These provide us with examples of mapping, unmapping, narrative mapping, anti-mapping or deep mapping practices that, like *Kaavad: Travelling Shrine: Home*, "interweave the historical with the contemporary, the political with the lyrical, the factual with the fictional, and the discursive with the embodied in ways that make unexpected connections".[34] Given this range we can only offer an account of those small portions of this ongoing and open-ended project that we can relate to directly through our own practices.

Towards an inclusive 'geopoetics'?

There are various strands of intellectual activity that anticipated elements of deep mapping/slow residency. These include the 'ethno-poetics' of the poet- anthropologist Gary Snyder, Robert Frodeman's work on 'field science' and interdisciplinarity as a solution to the insufficiency of specialised knowledge and the 'geopoetics' of Kenneth White.[35] Taken together, these strands provide one background context for our concerns. However, they need to be understood within the wider shifts in understanding place following the Second World War.

There are any number of individuals who have played an important part culturally in laying the ground for our concerns here. One is the Cornish-based English polymath Peter Lanyon who, by 1964, had established a position that, in certain respects, anticipates aspects of deep mapping. Lanyon's work is significant because through it he developed a specific, regionally based, painting practice that involved an intimate experiential relationship with place understood in depth and from a variety of perspectives.[36] This involved phenomenological engagement with a region of Cornwall through walking, bicycling, driving and flying. This multi-perspectival approach enabled Lanyon to move landscape painting on from its traditional fixed viewpoint. His commitment to place and place-ness required him to engage not only with the 'visual definition' of a 'specific location', but with 'its symbolic, social and historical association', its human as much as its non-human aspects.[37] Lanyon insisted that the tradition landscape concerns with nature and the countryside needed to be replaced with concerns about environment, place and the making visible of a time process. He hoped that the new landscape painting, like the sciences, would provide a clear sense of extra-human forces at work, a matrix in which landscape appears as just one, albeit diverse, component of an environment that cannot be isolated from the historical vicissitudes of human activity.

In 1975 the Scottish poet, academic and writer Kenneth White published *The Tribal Dharma: An Essay on the Work of Gary Snyder*.[38] This text is indicative of an emergent debate, published at a time when Snyder, an American poet, anthropologist and environmental activist, was assembling the ideas for his essay *The Politics of Ethnopoetics*. Snyder proposed a 'new Humanism' in which the blending of politics, anthropological investigation into the culture and ecology of non-literate peoples, and poetry are proposed as 'a new academic field'.[39] In the late 1970s White, who had also read extensively in anthropology and ethnology, had begun to use the term 'geopoetics'. He did so for two reasons. First, he saw that the biosphere was under threat and that it was necessary to develop new and effective ways of thinking and acting to protect it. Second, this insight coincided with his view of poetics, namely that a rich poetics depended on contact with and a reading of, the earth and biospheric space. In 1989 White founded the *International Institute of Geopoetics* to promote research into the cross-cultural, trans-disciplinary field of study that he had identified over the previous decade. White sees geopoetics as a praxis intended to renew culture by going back to the fundamental relationship between the human intellect and what, broadly speaking, is called the environment. He stresses, however, that this involves:

> more than 'poetry concerned with the environment', more than literature with some kind of geographical context . . . Geopoetics is concerned, fundamentally, with a relationship to the earth and with the opening of a world.[40]

There are both convergences with, and differences between, Kenneth White's understanding of geopoetics and our own. We broadly share his view that geopoetics can provide:

not only a place, and this is proving more and more necessary, where poetry, thought and science can come together, in a climate of reciprocal inspiration, but a place where all kinds of specific disciplines can converge, once they are ready to leave over-restricted frameworks and enter into global (cosmological, cosmopoetic) space.[41]

However, we would distance ourselves from certain aspects of his approach. Why can be indicated by referencing White's interest in the old Borders ballads like *Thomas the Rhymer* and *Tam Lin*, identified as an 'obstinate survival' of shamanic culture in Scotland.[42] For White, the importance of this survival is ultimately that it validates his sense of himself as a shaman without a tribe; as "a figure with a poetic-therapeutic role in society whose practice goes way beyond the role of 'the artist' in modern society".[43] This is in marked contrast to my own sense of the value of these same ballads (see Figure 1.3). While I also recognise the importance of the survival of polytheistic culture, my focus has been on the various popular uses made of these ballads, both historically by a female 'counter-culture' and as a continuing and vital medium through which contemporary individuals can renegotiate their identity.[44] This difference relates to wider issues, not least White's dismissive attitude to popular culture and spirituality. We find this aspect of White's geopoetics problematic, as we do his dismissal of the concepts of the region, regionalism and the realities of 'everyday ordinariness' in favour of the cosmic.[45] These differences lead us to turn from White to Robert Frodeman.

Figure 1.3 Gary Peters and Iain Biggs, *8 Lost Songs*, Artist's book, published by Making Space Publications in an edition of 50, 16 pages, CD of music, printed silk map and 8 loose printed images. Photo: Iain Biggs.

Before commenting on Robert Frodeman's work, we would add that the concerns of White's geopoetics were anticipated or echoed in different degrees by a variety of other significant thinkers, poets and visual artists. White himself tacitly acknowledges the poet St-John Perse and the painter Paul Klee as anticipating his own 'cosmological poetics'.[46] (Significantly, perhaps, he makes no mention of the German sculptor, political and environmental activist Joseph Beuys, whose own carefully considered use of a quasi-shamanic perspective offers a more nuanced alternative to White's own shamanic aspirations). Mike Pearson and Michael Shanks' insistence that their theatre/archaeology is *both* science and art perhaps echoes Frodeman's concerns rather than White's, while the work of the Canadian poet Don McKay responds to our living in a time of environmental crisis by intertwining concerns across geology, ornithology and the ethics of nature poetry; insisting that wilderness is best seen "not just a set of endangered spaces, but the capacity of all things to elude the mind's appropriations".[47] In short, we are ultimately responding to a collective set of concerns rather than a unique position.

Robert Frodeman

Much of Kenneth White's project for geopoetics is deliberately exalted, partisan and polemic. It promotes, that is, an exclusive vision framed in ways that limit its collective potential. Robert Frodeman's *Geo-Logic: Breaking Ground between Philosophy and the Earth Sciences* is, by contrast, a sober and practical argument for 'geopoetry, geopolitics, and geotheology' as sites to "disrupt the categories that have governed western culture since the birth of the modern age".[48] Frodeman works from an understanding that: "we are not making good use of our intellectual resources, in large part because of the disciplinary presuppositions that dominate the production of knowledge today".[49] In this and other respects we are wholly in sympathy with his approach. For example, we share his view that the prohibition within the field sciences (and many other academic disciplines) of what Michael Polanyi calls 'personal knowledge' is deeply counter-productive. It cuts scientists off from "a major source of their understanding, the intuitive grasp that comes from years of working intimately with a subject".[50] Or, in the terms we use here, cuts individuals off from the understanding absorbed through slow residency. The importance of fieldwork, and of this intuitive understanding more generally, is something we will discuss in detail in the final chapter.

Given what we share with Robert Frodeman's position, we might be expected to accept and develop it. Our difficulty here lies with his uncritical identification with philosophy and the earth sciences. This means that, despite his criticism of the disciplinary mentality, he still takes for granted some of the presuppositions and resulting limitations of the dominant logocratic culture we wish to move away from. We also have concerns about his particular conception of the sacred as implying "the centrality of the notion

Figure 1.4 Ilana Halperin, 'Boiling Milk, Solfataras 2000', photograph of the artist
actually boiling milk in a thermal hot spring. Her work typically merges
an interest in overlaying geologic formations and epic time with present
actions. Photo: Ilana Halperin.

of care – the capacity to restrain oneself, to control one's desires in order to
make space for another" while seeming entirely worthy, is tacitly framed by
presuppositions that we see as problematic.[51] It assumes, that is, an authoritative 'I' that restrains, controls and makes space. Indeed, Frodeman argues
that this same monolithic 'I', in the person of the earth scientist, should have
"the central role" in defining "the limits that individuals and communities"
must accept "to live within in order to flourish".[52] From our perspective such
presuppositions and privileging of one knowledge perspective over all others
are profoundly problematic. To indicate why requires us to return briefly to
our emphasis on a compound or ensemble self (Figure 1.4).

   The post-Jungian thinker James Hillman provides a detailed critique of the
presuppositions that, in our view, underwrite Frodeman's assumptions about
the self as the authoritative 'I' that paternally restrains, controls and makes
space. Hillman proposes a notion of self that is, by contrast, an "the internali-
zation of community" or what we have called a compound or constelled self;
a self-as- citizen imagined as an on-going, multi-dimensional conversation; as
"constituted of communal contingencies" rather "an immanent soul-spark of
a transcendent God".[53] From Hillman's perspective, then, Frodeman's 'I' is
the product of an "omnipotent fantasy" that replicates: "the splendid isolation
of the colonial administrator, the captain of industry and the continental academic in his ivory tower".[54] Why this matters practically can be illustrated by

referencing Frodeman's understanding of Paul Cézanne's paintings of Mont Sainte-Victoire; his emphasis on 'the act of envisaging', which he rightly opposes to that of 'collecting data', but which he still sees as concerned with an: "eye that has learned how to penetrate the object – probing it, testing it, sizing it up".[55] For us, this language again points to a significant difference between certain traditional presuppositions and our own concerns.

Our wariness regarding the instrumental tenor of 'probing', 'testing' and 'sizing up' returns us, in the context of the active ocular orientation Frodeman advocates, to our rejection of the perspective of "a culture hierarchized by a *logos* that knows how to speak but not to listen".[56] While we can find much in common with Frodeman's position, we are committed to being "apprentices of listening", rather than to laying claim to the mastery of discourse we detect in Frodeman's language of 'probing', 'testing' and 'sizing up'.[57] A language that ultimately belongs to a: "rationality which is expected to triumph *over* nature, which can speak but not listen in the same way and thus loses genuine interest in what it deals with".[58] (Our concern with listening is developed further in Chapter 2.)

Having scoped some of the possibilities and differences prompted by considering the geopoetics of thinkers such as White and Frodeman, both of whom have valuable insights to offer, it is now time to move on.

Notes

1 Felix Guattari, *The Three Ecologies* (London: Continuum, 2008).
2 Paolo Freire, *Pedagogy of the Oppressed*, trans. by Myra Raomos (New York: Bloomsbury, 1970).
3 Johanna Drucker, *Sweet Dreams: Contemporary Art and Complicity* (Chicago, IL: University of Chicago Press, 2003), 17.
4 Edward S. Casey, *Getting Back into Place. Towards a Renewed Understanding of the Place-World* (Bloomington: Indiana University Press, 1993), 31.
5 See, for example: Kenneth Frampton, 'Place-Form and Cultural Identity', in *Design after Modernism*, ed. by John Thackara (London: Thames & Hudson, 1988); and Carlo Ratti and Others, 'The Power of Networks: Beyond Critical Regionalism', *The Architectural Review* (23 July, 2013).
6 Doreen Massey, *For Space* (London: SAGE Publications, 2005), 9.
7 Tim Ingold, *The Perception of the Environment: Essays in Livelihood, Dwelling and Skill* (London: Routledge, 2000), 201. Tim Ingold, *Being Alive: Essays on Movement, Knowledge and Description* (London: Routledge, 2011), 71.
8 Our concerns as they relate to democracy can be understood in the following context:

> Most people idealise democracy as a form of government in which an informed populace deliberates about the common good and carefully selects leaders who carry out their preferences. By that standard the world has never had a democracy. Political scientists . . . who study how democracies really work are repeatedly astonished by the shallowness and incoherence of people's political beliefs and the tenuous connection of those beliefs to their votes.
>
> (Steven Pinker, 'Stranger Than Fiction', *Guardian Review* (13 August 2016), 3)

9 Paul A. Bové, *Mastering Discourse: The Politics of Intellectual Culture* (Durham, NC: Duke University Press, 1992), 3.

10 Barbara Bender, as quoted in Doreen Massey, 'Landscape as a Provocation: Reflections on Moving Mountains', *Journal of Material Culture*, 11, no. 33 (2006), 34. She wrote: "Landscapes refuse to be disciplined. They make a mockery of the oppositions that we create between time [History] and space [Geography], or between nature [Science] and culture [Social Anthropology]".

11 Ivan Illich, *Disabling Professions* (London: Marion Boyers, 1977), 11–40.

12 Geraldine Finn, *Why Althusser Killed His Wife: Essays on Discourse and Violence* (Atlantic Highlands, NJ: Humanities Press, 1996), 156.

13 Miwon Kwon, *One Place After Another: Site Specific Art and Locational Identity* (Cambridge: The MIT Press, 2004), 156.

14 Kwon, *One Place*, 161.

15 Ibid., 164.

16 Sally James, personal communication with Iain Biggs.

17 Francisco Pereria Da Silva, in conversation with Mary Modeen, December 2018.

18 For discussions of the continuing impact of the secular monotheism that continues to frame thought modelled on scientific analytics see: Ernest Gellner, *Postmodernism, Reason and Religion* (London: Routledge, 1992), 94–95; Zygmunt Bauman and Leonidas Donskis, *Moral Blindness: The Loss of Sensitivity in Liquid Modernity* (Cambridge, MA: Polity Press, 2013), 19–29; and the work of James Hillman and others engaged in a post-Jungian 'polytheistic' psychology.

19 Alan Garner, *The Voice That Thunders* (London: The Harvill Press, 1997), 41.

20 Owain Jones and Paul Cloke, *Tree Cultures: The Place of Trees and Trees in Their Place* (London: Bloomsbury Academic, 2002), 6.

21 Jones and Cloke, *Tree Cultures*, 10.

22 Ingold, *Being Alive*, 63.

23 Timothy Morton, *The Ecological Thought* (Cambridge, MA: Harvard University Press, 2010), 14.

24 Morton, *Ecological Thought*, 15.

25 Mary Modeen, 'Love from a Distance', paper presented at *Catchment/PLaCE/Mapping Spectral Traces* workshop, Bristol, 2011.

26 James Hillman, *The Thought of the Heart* (Dallas, TX: Spring Publications, 1984).

27 Garner, *The Voice That Thunders*, 41.

28 Finn, *Why Althusser Killed His Wife*, 137.

29 See, for example, Katherine Harmon, *You Are Here: Personal Geographies and Other Maps of the Imagination* (New York: Princeton Architectural Press, 2004), 49–51.

30 While we try to avoid the too-facile classifications of East and West as binary, overly generalised and sometimes indicative of essentialist thinking, we also have introduced these terms of reference occasionally when they are productive of a cultural contrast, or in allowing a space for comparative associations.

31 Howard Morphy, *Aboriginal Art* (London: Phaidon, 1998), 103.

32 Morphy, *Aboriginal Art*, 207.

33 Karen Till, *Mapping Spectral Traces: Exhibition Publication* (Blacksburg: Virginia Tech College of Architecture and Urban Studies, 2010), 3.

34 Till, *Mapping*, 3.

35 American poet-environmentalist, Gary Snyder (b. 1930–); American scholar of interdisciplinary studies with philosophy and geo-ethics, Robert Frodeman; and Scottish poet and academic, Kenneth White (b. 1938–).

36 Andrew Causey, *Peter Lanyon: Modernism and the Land* (London: Reaktion Books, 2006), 8–9.

37 Chris Stephens, *Peter Lanyon: At the Edge of Landscape* (London: 21 Publishing, 2000), 3–4.

38 Kenneth White, *The Tribal Dharma: An Essay on the Work of Gary Snyder* (Dyfed: Unicorn Press, 1975).

39 Gary Snyder, 'The Politics of Ethnopoetics', paper based on a talk given at the *Ethnopoetics* conference at the University of Wisconsin (April 1975), <http://angg.twu.net/LATEX/poep.pdf> [accessed 24 November 2017].

40 Kenneth White, *Geopoetics: Place, Culture, World* (Glasgow: Alba Books, 2003), 20.

41 Kenneth White, 'The International Institute of Geopoetics: Inaugural Text', *Scottish Centre for Geopoetics* (1989) <htttp://www.geopoetics.org.uk/what-is-geopoetics/> [accessed 24 November 2017].

42 Tony McManus, *The Radical Field: Kenneth White and Geopoetics* (Dingwall: Sandstone, 2007), 70–71.

43 McManus, *Radical Field*, 73.

44 See Iain Biggs, *Between Carterhaugh and Tamshiel Rig: A Borderline Episode* (Bristol: Wild Conversations Press for TRACE, 2004).

45 For a useful discussion of this second point see Kathryn A. Burnett, 'Place Apart: Scotland's North as a Cultural Industry of Margins', in *Relate North: Culture. Community and Communication*, ed. by Timo Jokela and Glen Coutts (Rovaniemi: Lapland University Press, 2017), 71–72.

46 White, *Geopoetics*, 32–33.

47 Don McKay, *Vis-à-vis: Fieldnotes on Poetry & Wilderness* (Wolfville: Gaspereau Press, 2001), 21.

48 Frodeman, *Geo-Logic: Breaking Ground between Philosophy and the Earth Sciences* (Albany: State University of New York Press, 2003), 2.

49 Frodeman, *Geo-Logic*, 16.

50 Ibid., 35.

51 Ibid., 55.

52 Ibid., 117.

53 James Hillman, 'Man Is by Nature a Political Animal or Patient as Citizen', in *Speculations after Freud: Psychoanalysis, Philosophy and Culture*, ed. by Sonu Shamdasani and Michael Münchow (London: Routledge, 1994), 35.

54 Hillman, 'Man Is By Nature', 33.

55 Ibid., 113.

56 Gemma Corradi Fiumara, *The Other Side of Language: A Philosophy of Listening*, trans. by Charles Lambert (London: Routledge, 1990), 85.

57 Fiumara, *The Other Side*, 57.

58 Ibid., 156–157.

2 Seeing, listening, acting

Taking an imaginary walk

Let's take a walk. The day is grey but not raining, the temperature not too cold. We need a breath of fresh air and to stretch our legs a bit. Perhaps you recall that in *The Twilight of the Gods,* Nietzsche wrote[1]: "It is only ideas gained from walking that have any worth". Walking as a practice, as a method of encounter and as an artistic mode of reflection and action has a long history: Rebecca Solnit's *Wanderlust: A History of Walking,*[2] Robert MacFarlane's *The Old Ways: A Journey in Foot,*[3] Roger Deakins' *Wildwood: A Journey Through Trees,*[4] Bill Bryson's *A Walk in the Woods,*[5] and Geoff Nicholson's *The Lost Art of Walking*[6] are a few among the many published works that revel in the action and consequences of walking, and all of these focus on a first-hand experiential encounter as the basis for their observations.

Walking artists – and artists who deploy walking as a strategy – are also many and well-established: Richard Long, Hamish Fulton, Francis Alÿs, Janet Cardiff and George Bures Miller, Phil Smith, the Situationists, and a great many others use walking as a central method of encountering phenomena and engaging with found places, situations and events, as many recent publications show.[7] As participants in a durational process, they often deploy strategic practices that unfold in interactive and chance-influenced narratives. Some of these artists intervene in the land they traverse, leaving something that alters the terrain (Long).[8] Others capture something from the walk, bringing it home to reflect further on the environment and moments of their experience (Cardiff).[9] More in-depth study in this area of walking practices, particularly as considered from the perspectives of human geographers and social anthropologists, is ably introduced by Lorimer in his essay 'Walking: New Forms and Spaces for Studies of Pedestrianism'.[10] The anthology *Walking, Landscape and Environment*[11] features many meditations prompted by the act of walking, and among these are aspects of theatricality,[12] philosophy and walking[13] and the women's walking library.[14] But oh dear, we have distracted ourselves before even taking a single step. So, mindful that this a process, an action, an exercise and a form of meditation or close attention shared by many, let us go together then, crossing the road and heading for the riverside. In the previous chapter the concept of 'a community of transverse action' was introduced as one of the terms presented. This was defined as a group of people who are engaged through their practices in events, listening and thinking separately and together, that cross disciplinary boundaries and, in our case, whose work involves practices in art and design, deep-mapping, academic research, teaching, writing, and various forms of social interventions, community actions, observations, fieldwork and eco-conservational activism. Even something as simple as a stroll can be in the first instance a chance to participate in active encounters. On a larger scale, these collective activities are driven by complexly interwoven experiences, questions and beliefs. First among these is the conviction that *to listen* and *observe* is to learn about the world we inhabit; *to imagine* is to create a new world through possibilities; *to speak* is to share understanding; *to make* art is to celebrate a world; and *to communicate* 'deep knowledge' is to draw on collective and intergenerational sited memory to rebalance life patterns that have skewed out of sustainable balance. It will also be apparent to you, dear reader, that there are distinctively different voices in this book. And this leads us to our second conviction: in order to see fully, and even to survive as a viable planet, certainly as a species dependent upon that planet, we not only *need* differences; we need to be open and permeable to those differences, valuing them precisely because they *are* different from our own, and give us a much wider understanding.

There is nothing new in this thought.[15] Different perspectives, different languages, different experiences, and a respect for multiple, overlaid perceptions result in a stronger, polyphonic approach to understanding our current world. Chances of species survival are based on differing gene pools, and different patterns of response to stimuli.

This same safety net plays out in ecosophical thought: thinking with 'two heads instead of one', or several heads, is preferable because the varying perceptions of the polyverse help us to arrive at different judgments and forms of action.

There is a third shared assumption that we authors have agreed and discuss in our own distinctive ways. This is the conviction that we are more than individuals, more than the monolithic 'I' implicit in Robert Frodeman's *Geo-Logic*,[16] which is bound in closed containers of skin. That is, we have come to believe in the ways that we extend beyond our individual 'selves', that we each are organic beings with energy, living agency and sensorial power that extends beyond singularity. It is a way of being in the world that allows for a kind of recognition of interconnectedness with other beings, human and non-human, and with the earth itself. Working with this belief, there is an imperative that is correlative: in attending to this 'overflowing' self, a self that is not a contained self, we are never solitary agents. Our actions have consequences for that to which we are connected. And this demands our attention to our environment, our fellow beings and the welfare of entire states-of-being. This is 'ecosophy'. Again, this is a reiteration of much that has been central to human wisdom for a very long time.[17] What is most important here is neither the very long tradition that lies behind a holistic attentiveness to the environment, nor any claims to a unique understanding of the self as permeable and fluid; rather, the significance of actions that are fundamentally interconnected to other people, beings and the earth brings a kind of foregrounded intercontextuality to the 'geognostic' actions discussed here, and to the dynamics of shifting relations. But as with every walk which must begin with one step at a time, as ours has done just now, let us begin with the individual, knowing that it is not *solely* the individual who is the focus of investigation.

Here is a simple statement: we humans attend to that which is not apparent. Spectral: beyond the visible, the audible, the signposted, or the accepted canon, we hear the rustlings of that which has gone previously undetected, that which is there but just under the level of the recorded or noted, off the map, for whatever reason. Along our way on this imagined walk, we see over there, those minute traces in the vegetation of previous habitation, for example, but more than with just the focussed intentions of a field anthropologist, or social historian, or human geographer, or poet who is seeing-at-while-seeing-past, or as the artist with an eye for detail; we aim to work with and between what these disciplines can show us, their methods of attending to the unseen. We humans, who are so good at construing meanings and patterns, of seeing and saying what we construe, sense that it is the trace of numinous presences, the pull of the land itself, the refusal of the earth to be still that lingers in our ears as the unheard sounds we know are there, the geophonies of the world itself, or the 'something' we thought we saw in the corner of our eyes, a shadow on the retina not made by passing clouds. As the poet Robert Bly wrote, "at last the quiet waters of the night will rise/And our skin shall see far off as it does under water".[18] What is this 'seeing far off' all about? Most of us are keenly familiar with the ability to know things that are difficult to say how we know. Is it instinct? Second sight? Sixth sense? Intuition? Neurocognitive pre-verbal thresholds? Very few words about the mechanisms of this ability convey its profound pervasiveness. The poet Don McKay, in speaking of his writing practice in a way that has a direct parallel with a studio practice, began by comparing the act of writing to one's mind-set as it occurs in birdwatching. He described this as "a kind of suspended expectancy, tools at the ready, full awareness that the creatures cannot be compelled to appear"[19] (Figure 2.1).

As we reside in this state of 'suspended expectancy' maybe 'being human' includes the conviction that we know in ways other than those most readily articulated. At least, it is often carefully phrased like this in the context of many Western contemporary mainstream academic circles. As we co-authors have written about before, singly and together, we need to look more at Indigenous cultures to find alternative ways of addressing this knowledge unashamedly and straightforwardly, or of honouring and

Figure 2.1 Norman Shaw, *Tir Nan Analog*, medium: ink and coloured pencil on paper, year: 2019, dimensions: 59.5 × 77.5cm. Photo: Norman Shaw. In listening deeply and sensing through deep attention, Norman draws on unseen but perceived energies to make visible what is not apparent to the eye.

accepting the fullness and deep mystery of the lived experience as quite normal. Whether through studying – for example the extraordinary aesthetics of *rasa* articulated by Abhinavagupta (c. 950–1020 CE) – or working in our research networks[20] with our colleague, Mona Smith, a Sisseton Wahpeton Dakota storyteller and media artist, and in Minnesota with her other Dakota friends during their anniversary year honouring their Dakota homeland, and with Ojibwe, Māori and First Nation friends and colleagues as well – we are reminded of the importance of remembering 'through the eyes of Indigenous reality', with the richness of other languages and art that fully presence us to the abiding presence of the unseen, configured in many ways, but always real. More on this will follow below, as well as in Chapter 9.

A word of caution is important here. One of the preliminary warnings taught to first-year students in universities is to watch out for the appeal to the reader or listener when you hear the word 'we'. Academics generally advise scepticism towards any spurious inclusiveness as the starting point for critical listening. It too securely assumes complicity, thereby compromising the full scope of objective critical analysis. It cosily sucks you, the listener/reader/fellow traveller, into a sense of comfortable shared purpose without full attention to the wayside, or recognition that you have been pulled in a direction that may be quite distant from you own. Nevertheless, as you read these chapters, you will note how often we authors keep using this general pronoun. It is not done with the sly purpose of misleading you, or circumnavigating conflicting opinion. In part, it is because it refers to the conversational process of writing to, for and across each other, knowing each other's ideas and works, hearing each other's voices in our 'inner ear' while we write, sharing editorial comments, and preserving alternate views as much as communicating unified visions. It is also this inclusion of you, our reader, whom we address with earnest resolve, serious conviction and empathetic collegiality but not as the uncritical friend. Our use of the 'we' is never unconsidered, nor duplicitous in its intent; rather it is rooted in our commitment to multiple voices, to embracing differences and to open invitation in discourse that we speak to you directly, and inclusively.

Our footsteps echo as we walk together: along our path let's take a small diversion. In ancient Greek theatre, the Chorus was a group who spoke in unison: their words were meant to be understood as comments on the action, on collective truths spoken by the many and not just the words of any one individual. The chorus was removed from the direct action of the narrative but was presented as a weight of consciousness and moral truths. Beyond consensus, the Chorus voiced truths too important to be heard as if by one speaker. Or even more pointedly, that any speaker had too individual a perspective on the action, and therefore was simply single in his or her subjectivity – lesser in wisdom because less able to draw on multiple perspectives than the voices of the collective. Since most ancient theatre-goers knew the dramatic plots well before attending the performance, the fullness of the audience's experience lay in seeing the actions unfold, almost as if by pre-destiny or fate, and hearing the bone-chilling wisdom of the Chorus as if hearing divine affirmation, through the mouths of many, transcending any one individual, and speaking in unison. There is a parallel here to the community of transverse action, discussed in Chapter 6, in that the voices are plural, greater than any one, and in the distinct voices and accents, across all the differences, fundamentals (in the musical sense) of thought, perception and action resonate, one amongst the other. Multiple voices speaking about a need for collaborative action cannot be ignored. But in preserving multiplicity perhaps there is a second reading of the function of the Chorus; the weight of many voices may also be unwelcome if one stands opposed to the views of the majority. If Antigone had not the courage of her own convictions, prepared to follow what she felt to be a moral imperative in defiance of the state, she would not have set herself in opposition to Creon's prohibition to bury her brother Polynices. Instead, as Sophocles has her, she defies the king's decree, positioning herself in a liminal state between ethical and legal actions, between the feminine as centre of the family and outside the family.[21] Her actions are those of one who positions herself against the multitude, and will not be swayed when driven by her own conscience or what she believes to be a moral imperative. The tragedy, as tragedy it is, lies in the inevitable consequences that she knows she brings upon herself by her actions.

Let us return to the main focus, the immediate path we have deviated from our walk, and consider as we amble along companionably that which we all truly share: childhood, for one. If we were fortunate as children, in our near or distant past, and whether we recall clearly or not, we spent some time looking at clouds, watching minnows at water's edge, playing in the sand or snow, digging forts, making tunnels, or simply feeling grass between our toes, actively engaging with our environment, wondering, shaping and exploring. We were 'taking in the more-than-human world' – one of our first teachers. We learned through our senses, through our whole bodies, as we felt, observed, heard and tasted what it was like to be in the world. In doing this, we laid down memories that have become, literally, a part of us.[22] Childhood experiences are key to the ways in which our brains work, and these experiences in themselves help to shape our physiological responses thereafter in life. Neural net pathways are developed in specific ways through the stimuli of early environments. More than producing memories, these shaping early years establish the exact nature of the physiological capabilities – and the distinctiveness – of how we will continue to function as sensate beings in the world as adults. [For many children now, these experiences are either lacking or mediated by more digital wonders in a kind of technological distancing, and with the effect of perhaps less direct experience in the out of doors than in previous generations. This is a real concern for how this will play out as this generation takes its place as the deciders of 'the fate of place'.[23] Will this distancing effect have a perilous consequence when the demands of an overtaxed and unbalanced planet require unflagging dedication and care? Will they have an attachment to the material 'stuff' of our environment?]

There remains the question of what we 'take in', how we as specimens of *homo sapiens* receive the sensory stimuli from the environment and how we process this information, in the belief that this is the first step in a conscious process of crucial attention to the world around us, beyond the ways that we are shaped by it. An artist, a philosopher, or a cognitive psychologist, each would describe this process of attending to the world, and 'making sense' of it, in different language. It is not so much that each of these disciplines is concerned with separate phenomenon (although to a certain extent they are!) – as that they use different methods to study perception.[24] If factors of cultural differences – such as history, beliefs, social groupings, patterns of life and work, geographical factors, weather, topography, agricultural land use, flora and fauna – are added as variations in *what* is read, then the resultant combination of perceptions, interpretations and languages to convey these understandings expand exponentially. Many philosophers, philologists, linguists and poets have written about the importance of language in permitting the conceptual framework to extend to nuances of perception that are denied in other languages. Some perceptions simply don't translate because they are so idiosyncratically significant to the individual's experience that they stand outside of conventional experience; finding the way to share what we *do* perceive is the challenge.

Acknowledging this discussion, we co-authors believe that this research also brings the opportunity to share both practical and philosophical aspects of how it is that we know what we think we know, how we variously perceive the world, how we remember experience, and how our cultural values affect our memories. As artists and practitioners, we have our own understandings and revelations that have come through our practices, augmented by an academic education, lifelong teaching and careers as researchers. What we see (and don't see) fascinates us. Invariably, our language will lend a different sound and shape to this discussion from that of, say, a cognitive neuropsychologist, for example, or a human geographer. (One reason why we have restricted the range of academic material we reference to those texts that relate directly to our own learning.) We have much to gain from those fields, and many others, as do scholars from those other disciplines from artists. In a multidisciplinary investigation, in 'communities of transverse action' as we will discuss later, there are many facets of this topic that reflect each other's focus, despite differences in methods and idiom.

But as another single step in our imaginary walk, let's briefly reconsider how we take in the world around us. We walk together, we talk, and we notice the trees, the blue skies with billowing white clouds, the smells of trees and plants around us. Peripatetic movement is a key part of exploring the environment we inhabit. Our perceptions are not solely the physiological activity of the eyes, nervous system and cerebral cortex. Rather, they are a rapid and complex set of 'negotiations' of sensory stimulus and multi-faceted interpretation; the latter constitutes aspects of subject-specific experience, education and expectation, of whole-body sensation[25] and of *culturally influenced values* and *philosophical beliefs*.[26] It is precisely this set of phenomenological filters that shapes the interpretation of the neurological impulses received in the brain.[27] Here is an example: I interpret the blue skies to signify continued warm weather and lack of precipitation, always a joy in Scotland I might add. But just as likely, I could have had other thoughts about changes in weather patterns, comparing this point in the calendar year to the alteration in seasonal conditions. My interpretation, and my subsequent delight and/or concern, is influenced by many factors and previous experiences. In short, as adults, we see largely what we expect to see.[28] This leads us straight to what Canadian poet Don McKay means when he defines 'wildness' as 'the capacity of things to elude the mind's appropriation'.[29]

McKay writes poems that hover in 'the traces in which rock and stone have been left in language'. What does it mean to be turned to stone?' "A-stonied", he says, or "astonished," making the link between humans and stone, between stillness and movement. He addresses the earth itself as much as us when he writes:

Astonished

astounded, astonied, astunned, stopped short
and turned toward stone, the moment
filling with its slow
stratified time. Standing there, your face
cratered by its gawk,
you might be the symbol signifying eon.
What are you, empty or pregnant? Somewhere
sediments accumulate on seabeds, seabeds
rear up into mountains, ammonites
fossilize into gems. Are you thinking
or being thought? Cities
as sand dunes, epics
as e-mail. Astonished
you are famous and anonymous, the border
washed out by so soft a thing as weather. Someone
inside you steps from the forest and across the beach
toward the nameless all-dissolving ocean.[30]

'The nameless all-dissolving ocean'. Only a man who knows erosion, who knows beaches, who knows the power of water and time stretching out of the vast and featureless expanse can write this. Right from the outset, we are discussing differences in culture: the learned influences, as cultural shapers, which are profound and pervasive. Cultural values are read in part as cultural signifiers. And yet even within this knowledge, McKay's poem opens out other possibilities, ways of being Other within our own reactions: 'someone/inside you steps from the forest', shifting from the known inhabited spaces with markers and milestones, to vast watery expanse . . . The multiplication of perception lies here in the 'someone' who inhabits one's self, one within the multitude of selves. It is this one-of-the-many who is complicit in the 'deep knowledge' of stone and water, in the experience of the darkness of the forest contrasted with the light of the clearings, and in understanding the surface waves covering over the unknown, the unexperienceable. Simultaneously, McKay's appeal is both inwards (to the one-of-the-many, the Other within) and to the outward-facing 'all-dissolving ocean'.

Let us consider an example of another type of 'layered knowledge' offered by the critic of Orientalism, the late Edward Said, who suggests that Jerusalem is a place of cultural overlays:

a city, an idea, an entire history, and of course, a specifiable geographical locale often typified by a photograph of the Dome of the Rock, the city walls, and the surrounding houses seen from the Mount of Olives; it too is overdetermined when it comes to memory, as well as all sorts of invented histories, all of them emanating from it, but most of them in conflict with each other. This conflict is intensified by Jerusalem's mythological – as opposed to actual geographical – location, in which landscape, buildings, streets, and the like are overlain and, I would say, even covered entirely with symbolic associations totally obscuring the existential reality of what as a city and real place Jerusalem is.[31]

A city is both a place *and* a meeting of cultural signifiers – and for more than one culture in this case, Jerusalem is symbolic of existence itself. It is an overlay of experience and history, of personal and cultural, religious and secular experience. If we point to the space between what Said calls 'the existential reality' of the city, and what it has come to signify, we might be tempted to use Derrida's term, *différance*,[32] meaning not only 'difference' but the space of a deferred, always-unfolding process of signification, knowable only by privileging the visual. It is a process of meaning coming into play as the reading of the place is 'read' by each visitor, understood in its dissimilarity in each and every instance of its confrontation.

Let us extend this attention to difference. For example, to many or even most Indigenous communities around the world, culturally based attunements to the invisible presences and traces of the unseen are accepted and known as parts of daily life. They are fundamental beliefs within these cultures, lying at the centre of their social values, forming the core attitudes of the relationships of the individual to the earth itself.

As individuals, one of the first things we begin to learn about is 'our place' in this world, in a continuous process throughout our lives as we discern what this place is, and how we accept or resist or decide to change this locus we inhabit. As adolescents or young adults, we extend this experience to temporal concepts: mortality is the ultimate map, defining the parameters not just of geographic place, but of existential place. *Hic Jacet* means 'here lies', a frequent epitaph on tombstones. As Harrison notes in his essay *Hic Jacet*,[33] grave markers are the first signs (*sema* in Greek, the same word for both 'grave' and 'sign'), marking the 'here' of human mortality.

Remembering, memorialising is marking the place on earth in human terms. Harrison writes,

> It is as if we the living can stand (culturally, institutionally, economically, in other words humanly) only because the dead underlie the ground on which we take our stance.[34]

For Harrison, place *as place* is marked by human consciousness; human actions; and, most of all, by human mortality. 'Placelessness', if we construe from this line of reasoning, is therefore un-demarcated; land-wildness is place without boundaries. He eloquently continues:

> For what is a place if not its memory of itself – a place where time reflects back on itself? The grave marks a site in the landscape where time

cannot merely pass through, or pass over. Time must now gather around the *sema* and mortalize itself. It is this mortalization of time that gives place its articulated boundaries, distinguishing it from the infinity of homogenous space.[35]

Phenomenologists and the cognitive scientists of today would argue that embodied perception – the *lived experience* – is the first way we truly know the world in the sense that our bodily senses inform us of what surrounds us. By definition 'placelessness' cannot be experienced bodily or temporally, then, and can only be a construct of the imagination.[36]

Poets would agree, basically, but use the language of their own experience to evoke the knowledge of 'body-memory', if we can call it that. Poststructuralist philosophers would say that it is the fact that our own 'always already' being[37] is actually double: in other words, our bodies as both object (*corpus*- that which contains our sensory apparatus) and subject (the conscious embodied-being who feels ourselves being touched). We can touch our own hand with the other, sensing the touch while simultaneously feeling the touch of the other hand. And second, because we cannot think back to a time when this was not the case – the simultaneity of sensing the world is being aware of being the Sensor. Simple. So why is this important?

History, records, maps and charts are the substance of retrieval, but what is retrieved is the account of what was 'grouped' (or collected) in the first place. If the process of 'making sense' of an event is told by one person, then stored, and then later retrieved, all we can know of the event is the singular 'taking in' and the single narrative retold. This is unreliable evidence, forensically speaking, however much we enjoy the storytelling. Singularity of perspective stands in opposition to the very process of multiple perceptions we have already discussed. It might be history, but in historical accounts it is always important to ask *whose* history this is. The events that defy 'making sense' – or the imperfectly formed memories of those to whom the events can't make sense – or even the missing accounts whose memories are not stored, or are devalued, are equally history, but may not be retrieved.

There is one last thought we must pursue before we end our walk, and that is this idea of 'porosity'. Walter Benjamin suggested that the way that we know things often relies upon a language or visualisation of containment. Experiencing a city, for example, as in our example of Jerusalem above – or in Benjamin's case, in Naples[38] – one walks through neighbourhoods where there is an identifiable character, and we note its flavour or distinctiveness in making sense of where we are. But there are 'spatial porosities' he suggests, and these are instances where there is an overlapping of characters, or a permeation of barriers and boundaries. In fact, in these instances, it is difficult to determine boundaries at all, and instead experientially tend to note qualities by *transitions* instead of boundaries. This is a useful framework for perception, relying less on set definitions or closure, and more on a process of ongoing and nuanced observation system. It has an equally compelling

Figure 2.2 Falls of Bruar, Perthshire, Scotland. The shifting striations of rock which
are remnants of vast volcanic forces, stone arches, dripping vegetation
and the drop to the water below create this layered place. Photo: Mary
Modeen, 2016.

counterpart in current identity theory, for example, where there is less emphasis placed on fixed characteristics, and shift toward more instances of
recognition (Figure 2.2).[39]

This interests us greatly because many individual projects, and those of
the authors in particular, work as artworks built upon this sense of porosity or permeability. Aware of inhabiting spaces that are both physical and
metaphysical, where transitions and overlaps abound, and even the point of
view is permeable, we work with multiple and shifting elements, so that fixity slips away as every notional vision elides with memory. It is indeed uncertain territory, akin to the different river we step into with each immersion.

The 'Othering' of perception

These various accounts, what we might call 'minority reports', are essential
to history and historical processes: first, for the same reason given above,

that multiple perceptions are the core of human experience. But also these recollections are unreliable unless substantiation from other perspectives is offered. This has been the basis of feminist scholarship and Indigenous studies for the last 30 years or more. The Othering of history is the recovery of difference; the restitution of alternative narrative, and more specifically the nuancing of narrative, is the process of valuing multiple perspectives. Quite literally, this is a process to reject singular recall and to replace cultural forgetting with cultural remembering, quintessentially with multiple perspectives, with the fullness of emotion, with the languages of the separate classes, with the multi-sensory and multi-modal, and with the ways in which individuals feel real events.

Richard Rorty in his book *Philosophy and the Mirror of Nature*, suggests that much of what passes as the questions philosophers ask are what he calls 'bad questions'. They are 'bad questions' because they fall out the realm or ability of philosophers to deal with them in any meaningful way. He suggests that analytical philosophy has been too often a linguistic game of 'pseudo–problems'; one group of philosophers poses a question, and by meandering through experiments and language problems, the resultant work is years of intellectual adaptation of new definitions until no one remembers the original question any longer and another question is asked. To think about the nature of reality, for example, he says is too big a question to ponder meaningfully. He asks, if we learn that world is to be imminently destroyed by meteors, should philosophers begin to contemplate meteors? No, he concludes, there is no point. There is nothing they can add, change, do or produce. The bad questions should be 'dissolved'.[40] In this pragmatist approach, or 'neo-pragmatist' as some have termed it, there is no one privileged view of nature as a representation or an interpretation, that by 'digging through' one is able to arrive at a 'truth' (Figure 2.3).

But what can we make of this? Are we, as artists and writers and thinkers, absolved from anything other than making art, poems, stories or thinking? What is our responsibility as thinking, moral creative souls who wish – as Thoreau would say – 'to live deliberately'? Perhaps an artist concerning her- or himself thoughtfully with the environment is not the same thing as a philosopher contemplating meteors that will plunge into the earth. There may be several reasons for this. First, we all know that the planet has been suffering, that the danger has been human-made in origin, and that it requires immediate and concerted action if we are not to destroy ourselves as a species and most of the world in the process (Figure 2.4).

In a long line of literature, including Rachel Carson's *Silent Spring* (1962) and Paul R. Ehrlich's (2013) review of his own 1968 book *The Population Bomb*, the wake-up call has been stark and clear. Their consensus, bluntly stated, is that humans are failing in the stewardship of the planet, growing too numerous, killing ourselves and millions of other species, and poisoning the Earth. How can we artists, writers and thinkers ignore this? Is this our most important long-term task – to use our art to encourage consciousness,

Figure 2.3 Phil Braham, *Forest Entrance, Bridge of Orchy* from the *Suicide Notes*, photograph, 2009. In this series of photographs, Braham visits the sites of previous suicides, looking for the 'tell' of past events, or evoking unseen associations, challenging the nature of representation itself. Photo: the artist.

to increase awareness and to prompt concerted action? This is not 'betraying our calling as artists' by distorting art into propaganda, as some would have it. No; moral conscience does not equate with propaganda. Donna Haraway chides us to keep with the problem, to continue to put our actions and minds to and toward a reconstruction of how humans can live in the environment and with all other creatures, and prompts us to be creative.[41]

Effectively, it is a task of creative 'ecosophy', engaging in actions with the informed knowledge of an ecological philosophy. Through communities of transverse action, it may be possible to resurrect a hope that mitigates against the despair, bludgeoning of dire forecasts and unrelenting evidence of mistakes, and thereby to share empathetically a deep knowledge of how the Earth itself bears the scars of ignorant actions.

Let's suggest a kind of ignorant inattention by way of contrast in order to understand the problem from a different angle. Benjamin theorised in the *Little History of Photography*[42] on the differences between the camera and the human; he introduced the notion of photography as revealing the world in the form of the 'optical unconscious' [*Optisch-Unbewussten*]. This is quite distinct from Freud's psychoanalytical unconscious, for example. In Benjamin's understanding, the camera mechanically depicted but in itself

Figure 2.4 Laura Donkers, *Local Food for Local People*, Socially engaged community action on the Hebridean island of North Uist, 2015–18. Photo: the artist.

did not rise to a level of cognisance, much in the same way that is possible for human and non-human animals to 'see' something as a process of visual sensation but not to be able to make it signify. It is a viewing at a removed distance, and not necessarily attendant with meaning. In human subjects, this effect might happen inadvertently when presented with something that is entirely unfamiliar. Or it might happen by an effort of will, when the interpretive act is suspended or repressed, such as in moments of duress or in altered states of consciousness. It seems that Benjamin's *optical unconscious* has a parallel with an 'ecological unconscious', so to speak, whereby the visible evidence of ecological damage and destruction is evident, but the population does not signify on a cognisant, active level. In short, there is a disconnection between sensation, the 'taking in' of information and the ensuing lack (apparently) of understanding, of horror, or of remedial actions.

Doom and gloom. Disaster. 'Sure and certain'[43] destruction. We've all heard these words and had these thoughts countless times.

Depressing, isn't it? But do we change our lives? Perhaps we are the rabbits, frozen into catatonic rigidity as the headlights approach. Or perhaps we suffer from 'nature deficit disorder' as has been articulated by ecologist David Siegenthaler.[44] His writing signals that hope, wonder and firsthand knowledge may be the best possible steps towards positive ecological actions. Simply taking a rural walk outdoors with a child opens one's eyes to the day. Siegenthaler suggests:

> One source of hope for deep ecologists, in the struggle to effect radical reform, is similar to the hope held by those of various faiths: that it is still possible to experience the wonder and grace of a world that, despite all the hurt, is still marvelous and full of possibility.

So yes, knowledge and experience can lead to wonder, wonder to love and love to hope. If we fashion a plan of action for the planet and for ourselves that is based on love, we may just yet endure. It is not very fashionable for an academic to speak of love, or an academic and artist to speak of a desperate need for our work to address hope. *But unless we all speak and act with courage –* and clarify that we are not speaking about the soppy sentimentality that passes for romantic 'love' in popular media – we all have little time left to be anything, much less specifically or exclusively artists or writers. We must share Antigone's courage to recognise an ethics that is 'pre-political' but absolute.

Earlier in our walk, we spoke of the importance of Othering, and particularly of the Othering of perception. We do not see the same things, we do not inhabit the same intellectual or spiritual spaces, we do not use the same colours to visualise the world's myriad phenomena. If our project is to succeed, we *need* differences just as surely as we need both air *and* water. It is more than respecting the differences of others; it is a positive need to speak with polyphonic voices, humbly respectful of differences, but in concert (Figure 2.5).

Invariably, these multiple voices are inflected with accents, with different perspectives and with emotions. Louise Glück, the American poet, writes with a mournful air linking life with loss and change as ways of knowing, in her poem about cottonmouth snakes and their territory:

Cottonmouth Country

Fish bones walked the waves off Hatteras.
And there were other signs
That Death wooed us, by water, wooed us
By land: among the pines
An uncurled cottonmouth that rolled on moss
Reared in the polluted air.
Birth, not death, is the hard loss.
I know. I also left a skin there.[45]

Glück's reference to birth as the arrival of a new self, another stage of being in which the emergence of a greater presence – greater for its closer attention to 'signs' – is also at a cost, and yet brings a kind of grief for what has been

Figure 2.5 Arthur Watson, *Memoria 2 Realities*, work-in-progress: studio shot, woodcut with carved letter text (2013). Watson links the local words for qualities of snow in a 'language' that bridges place, vision and text, but is nonetheless quintessentially Scottish. Photo: Mary Modeen.

left behind. In a speculative parallel process, if we humans manage to address and reverse the ecological damage of an unsustainable life economy, there will undoubtedly be a similar kind of grieving for a lost way of life that will be gone forever – no matter how successful and balanced the new patterns might be. As Glück says, "Birth, not death, is the hard loss".

We are convinced that it is imperative to say what we believe; as we walk with you and write for you, for fellow geognostics and the benefit of all living things. What is proposed here is an amended, alternative approach to what we do as artists, as writers and especially as teachers, parents and role models. It is this which echoes the beginning of this chapter: to speak is to create a world; to make art is to celebrate a world, through the materials of primary engagement; to imagine is to create a new world; and to share knowledge is sorely needed to rebalance life patterns. In our community of transverse action let us believe that we are acting together on our creative imaginations with a degree of urgency and importance, and that it is not just for ourselves.

People coming together define the word 'community'. They cohabit, and in near proximity, but more than that, they usually collaborate, at least on a partial level. Acting together, and acting with agency as individuals within a collective is the best way, perhaps the only way to acknowledge our attachment to the Earth, and to put embodied knowledge into a positive process. It is linking to deep knowledge, to the kind of knowing other than that which is easily explainable. As we walk, we see differently, we listen to each other's stories, and we act when speaking about what we know to be correct. In the broadest sense of knowledge, action must always be embedded within wisdom.

Notes

1 Friedrich Nietzsche, *Twilight of the Idols*, or *How to Philosophize with a Hammer* (1889), English version, trans. by R.J. Hollingdale (New York: Penguin, 2003).
2 Rebecca Solnit, *Wanderlust: A History of Walking* (London: Verso, 2000).
3 Robert MacFarlane, *The Old Ways: A Journey on Foot* (London: Hamish Hamilton, 2012).
4 Roger Deakin, *Wildwood: A Journey through Trees* (London: Hamish Hamilton, 2007).
5 Bill Bryson, *A Walk in the Woods: Rediscovering America on the Appalachian Trail* (New York: Broadway Books, 1997).
6 Geoff Nicholson, *The Lost Art of Walking: The History, Science, Philosophy, Literature, Theory and Practice of Pedestrianism* (Essex: Harbour Books (East) Ltd., 2011).
7 See, for example, David Borthwick, Pippa Marland and Anna Stenning, eds., *Walking, Landscape and Environment* (London: Routledge, 2020).
8 Richard Long's interventions in the land he has crossed are found in many examples. See: *A Line Made by Walking*, 1967, with documentation held in the Tate Britain, London.
9 Janet Cardiff and George Bures Miller often capture sound recordings of the ambient auditory environment, and then bring these back to the studio to work further with 'found' sound's ability to evoke embodied responses. See: *Lost in the Memory Palace: Janet Cardiff and George Bures Miller*, displayed at the Art Gallery of Ontario, April 6–August 18, 2013. <http://ago.ca/exhibitions/-lost-memory-palace-janet-cardiff-and-george-bures-miller> [accessed 10 April 2020].
10 Hayden Lorimer, 'Walking: New Forms and Spaces for Studies of Pedestrianism', in *Geographies of Mobilities: Practices, Spaces and Subjects*, ed. by Tim Cresswell and Peter Merriman (London: Routledge, 2010), 19–34.
11 David Borthwick, Pippa Marland and Anna Stenning, eds., *Walking, Landscape and Environment* (London: Routledge, 2020).
12 Cari Lavery, 'Walking and Theatricality: An Experiment', in *Walking, Landscape and Environment*, ed. by David Borthwick, Pippa Marland and Anna Stenning (London: Routledge, 2020), 36–50.
13 Frederic Gros, *A Philosophy of Walking*, trans. by John Howe (London: Verso, 2015).
14 Dee Heddon and Misha Myers, 'The Walking Library for Women', in *Walking, Landscape and Environment*, ed. by David Borthwick, Pippa Marland and Anna Stenning (London: Routledge, 2020), 113–126.
15 Darwin's evolutionary theory in *On the Origin of Species* (1859), for example, is based on valuing differencing, mutations and natural selection. This can be applied as much as to critical thinking as it can to evolutionary development, finding strength in alterity.

16 Robert Frodeman, *Geo-Logic: Breaking Ground between Philosophy and the Earth Sciences* (Albany: State University of New York Press, 2003).

17 Most Indigenous belief systems are quintessentially bound to the interrelationships between the earth and living beings. I would also include many of the world's great religions as well, in their invocation of 'right thinking' and 'right actions'. For example, see Bhikkhu Bodhi, *The Noble Eightfold Path: The Way to the End of Suffffering* (Kandy: Buddhist Publication Society, 1994). This is not to locate 'ecosophy' solely within religious or cosmological concerns, but to indicate that the scale of ecosophical thinking is central to the history of human understanding.

It is also important to note that groups who have recently appropriated the term 'indigenous' signify in very different ways than the origin of the word. For example, the British National Party embraces its membership with claims of Indigeneity: "Given current demographic trends, we, the indigenous British people, will become an ethnic minority in our own country. . . " British National Party, 'Stopping All Immigration', <http://www.bnp.org.uk/policies/immigration> [accessed 10 April 2020]. In other words, they use the term as code for 'white, English-speaking British-born residents'. Mindful of this caution in the use of the word, and its appropriation by xenophobic nationalists, I continue to write about 'Indigenous' peoples such as the Maori, the Native Americans, First Nation Canadians, Koori, etc. because there is no other suitable term that embraces these diverse peoples with ancient and evolutionary links to the land. To distinguish this specific use of the term for these peoples, I use a capital 'I' letter.

18 Robert Bly, 'Surprised By Evening', in *Silence in the Snowy Fields: Poems* (Middletown, CT: Wesleyan University Press, 1962), 15.

19 Don McKay, 'Some Remarks on Poetry and Poetic Attention', in *The Second MacMillan Anthology*, ed. by John Metcalf and Leon Rooke (Toronto: MacMillan, 1989), 206–208.

20 PLaCE International, Land², and Mapping Spectral Traces.

21 Judith Butler, as one of many authors who considers Antigone's actions, represents her actions as caught between but transcending kinship and duty, ethical rights and moral wrongs in a form of 'Pre-Political Consciousness', in Judith Butler, *Antigone's Claim: Kinship between Life and Death* (New York: Columbia University Press, 2002), 1–26.

22 Neuro-cognitive psychologists would describe several detailed processes of laying down memories through deposits of amino acids and then proceed to describe the complexities of memory retrieval through various neural net pathways. The point here is that memory has several component elements, involving encoding experience, storing memories and deploying a retrieval process.

23 There is a deliberate echo here of the title of Edward S. Casey's important book, *The Fate of Place: A Philosophical History* (Berkeley: University of California Press, 1997).

24 Here, this is *not* to say that perception and knowledge are synonymous!

25 Notably the work of Maurice Merleau-Ponty, *Phenomenology of Perception*, trans. by Colin Smith (New York: Humanities Press, 1962) and (London: Routledge and Kegan Paul, 1962) trans. rev. by Forrest Williams (1981; repr., 2002).

26 See Chapter 5: Polyvalent Perception and Cultural Values.

27 Interpretation is the key word here, and many philosophers and psychologists study this process by which we are able to attach significance to that which we sense. It is only in recent years that sensation and perception have become more distinct fields, as opposed to one in psychological study. And in philosophy, the entire concept of ascribing meaning is a core activity, and may be said to be central to the fields of hermeneutics, aesthetics and metaphysics, for example.

28 To a certain extent, psychologists refer to this phenomenon as 'confirmation bias'.

29 Don McKay, *Vis a Vis: Field Notes on Poetry and Wildness* (Wolfville: Gaspereau Press, 2001), 21.

30 "Astonished" from *STRIKE/SLIP* by Don McKay, Copyright © 2006 Don McKay. Reprinted by permission of McClelland & Stewart, a division of Penguin Random House Canada Limited. All rights reserved.

31 Edward Said, 'Invention, Memory and Place', *Critical Inquiry*, 26, no. 2 (Chicago, IL: University of Chicago Press, 2000), 180.

32 Derrida posits that the signified and the signifier have an asynchronous character and that this prevents any possibility of unity in the sign. The subsequent space between the two generates the precondition of the sign: that is, the deferral of language, referred to by Derrida as *différance*, simultaneously produces and denies meaning. As an additional note here, in this neologism, the spelling of the word is the same as 'difference', but the diacritical accent mark also privileges the sound of the word as being distinctive, and therefore signifying something different. Listening is critical here, too.

33 Robert Pogue Harrison, 'Hic Jacet', in *Landscapes of Power*, ed. by W.J.T. Mitchell (Chicago, IL: University of Chicago Press, 2002), 349–364.

34 Harrison, 'Hic Jacet', 353.

35 Ibid.

36 It is interesting to note the shift in attitude and definitions over time as described by Edward Relph in the period between two editions of his book. In the1970s he wrote:

> there is a geography of places, characterised by variety an meaning, and there is a placeless geography, a labyrinth of endless similarities. In this postmodern era, things are not so clear. . . My inclination now is to see landscapes not simply as revealing *either* place *or* placelessness, but everywhere as manifestations of both. . .

Edward Relph, *Place and Placelessness* (London: Sage Publications, 2008), preface to 2nd Edition.

37 Paul Ricoeur used this phrase first in his *Time and Narrative* (1983–85), 57. However, it is also relevant to Heidegger's construct of *Dasein* – literally 'being there' or being in place – which anticipates a state of being "ahead of itself".

38 Walter Benjamin and Asja Lacis, "Naples", in *Frankfurter Zeitung*, trans. by Edmund Jephcott (1925).

39 Paul Ricoeur, *The Course of Recognition*, trans. by David Pellauer (London: Harvard University Press, 2005), originally, *Parcours de laReconnaissance* (2004), 297.

40 Richard Rorty, *The Mirror of Nature* (Princeton, NJ: Princeton University Press, 1979).

41 Donna J. Haraway, *Staying with the Trouble: Making Kin in the Chthulucene* (Durham, NC: Duke University Press, 2016).

42 Walter Benjamin, 'Little History of Photography' ('Kleine Geschichte der Photographie', 1931), 371.

43 These words, spoken at the graveside, have always haunted me. Not only is their paradoxical relationship underscored, but when used in the funereal occasions are always linked with 'hope'. 'Sure and certain hope' seems to be not at all certain. Embracing uncertainty without giving in entirely to despair is also embedded in Haraway's work.

44 David Siegenthaler, 'Earth Walk: A Deep Ecology Perspective and Critique of the Mainstream Environmental Movement', *Unbound* (2012) <http://justiceunbound.org/journal/current-issue/earth-walk/> [accessed 9 November 2013].

45 Louise Glück, 'Cottonmouth Country', in *The First Five Books of Poems*, ed. by Louise Glück (Manchester: Carcanet, 2014). 41. Reprinted here by kind permission of Carcanet Press, Ltd. of Manchester, UK (UK and Commonwealth), and by kind permission of Harper Collins (US, Canada, Philippines and open market) 'Cottonmouth Country', in *Firstborn* (1968).

3 Deep mapping and slow residency

Part One: On *not* defining deep mapping/slow residency

Clifford McLucas, in thinking about the possibilities that deep mapping might offer, said:

> Whilst I can talk about deep maps, whilst I can imagine such things . . . whilst I can even dream about deep maps, unfortunately, I have to admit that I have never seen one.[1]

Clearly, he recognised that within deep mapping approaches there were both new possibilities for embracing diversity and contradictions, but also foresaw the risks inherent in defining an emergent technique too exclusively, with restrictive and overly defined parameters.

In parallel with McLucas, and framing this discussion, we will *not* begin with a definition of deep mapping/slow residency as a category of creative work. Anwen Jones and Rowan O'Neill see McLucas' observation above as a "pessimistic assessment of the practical realisation of deep mapping".[2] We prefer to think that McLucas understood that, in imagination and dreams, deep maps must always exceed our ability to realise them. Consequently, we prefer to see deep mapping less as a means to produce deep maps than as an impetus to a practical creative *mentalité*. *Not* providing a definition of deep mapping/slow residency is one way of keeping that impetus fluid within certain limits, of focussing on the complex interwoven-ness, relationality and blurring of divisions between diverse practices, and of holding at bay an overly restrictive definition of what is and isn't included. A term like deep mapping, taken non-normatively, brings with it a sense of the potential richness and possibility that imaginatively based acts of mapping might employ. The thinking behind this becomes clearer if we consider a specific approach to deep mapping/slow residency, as one example among a myriad of possibilities (Figure 3.1).

Silvia Loeffler, an artist, researcher and educator, reminds us that in Old Irish the word *glas* refers to all the shades of green, blue and silver of the sea, which she uses to evoke her concern with shades of place experience in

Figure 3.1 Silvia Loeffler, 'Transit Gateway: A Deep Mapping of Dublin Port'.
Photo: the artist.

her extended deep mapping project: the *Glas Journal*. Reading her two accounts of this project, with their emphasis on the personal (stressed through
the medium of artists' books), offers a very clear sense of her concern with:
'Tender and Chaotic Mappings'; of 'Liquid Mappings' that evolve into a
deep mapping.[3] This is a rigorous yet poetic approach, informed by visual
arts practices, a heartfelt feminism and a carefully developed consideration
of a harbour understood both as 'a Chronotopic Threshold Structure' in
the tradition of Bakhtin, and a prismatic evocation of those places that are
woven into our daily lives – those we name 'home', 'haven' or 'harbour'.

Working with more than 30 participants who agreed to document 'their'
place, Loeffler produced 28 hand-made, layered and hand-sized books that
evoke a series of harbour locations bordering the sea between the West and
the East Piers of Dún Laoghaire Harbour. In 2016 these books were displayed in the Maritime Museum using an installation form that echoes the
harbour arms and its metaphorical possibilities of containment, refuge and
protection in overlays.

Given different concerns, locations and their initiators' different backgrounds, deep mapping projects may have little in common beyond a sense
of their being an open-ended creative process deployed over an extended period. Each project draws out highly specific responses to equally particular
places as filtered through a unique mesh of interactions between that place,
the concerns, skills and understanding of the initiating individual, and
whatever her or his collaborators may contribute. To reduce the richness
and complexity of this multifaceted ecosophical interplay of environmental
and psychosocial relationships to fit a catch-all definition runs counter to

the particularism central to our understanding of deep mapping and slow residency. A definition that will require an analytical process of exclusion is likely to inhibit interest and investigation. However, lack of such a definition does not preclude rigour or value judgments.

Contrasting views

Deep mapping means different things to different people. By distancing ourselves from academic concern with 'deep maps . . . utilizing neo-geographic technologies', 'GSI and spatial analysis techniques' as 'a completely different, and radically new, method of exploring texts', or with 'the deep map as an access mechanism presenting aspects of several spatial narratives' is not to reject digital technologies or innovative programming.[4] Rather, it reflects a desire to keep at arm's length instrumental attempts by academics to make deep mapping serve restrictive disciplinary ends; to avoid conforming to an outmoded epistemology that continues:

> trajectories of enlightenment/modern aspirations of progress towards truth through the elimination of doubt and the application of reason, language and power in the dividing, sorting, representing and fixing of the world.[5]

There will, then, be a distinct and singular quality to any deep mapping/ slow residency project that genuinely responds to place.

This is reflected in the contrast between two publications on deep mapping: the special double-issue on deep mapping, later published as a book entitled *Deep Mapping*, produced by the e-journal *Humanities*, and *Deep Maps and Spatial Narratives (Spatial Humanities)*.[6] The first, edited by Les Roberts, reflects a view of deep mapping that, broadly speaking, recognises its entanglement in skills associated with the arts, alongside spatial anthropology, urban cultural studies, cultural memory and spatial humanities. Three American professors edit the second: Bodenhamer (History), Corrigan (History and Religion), and Harris (Geography), and frames deep mapping as a 'progressive' orientation within the Humanities. It attempts to use new technologies to reverse the Humanities' loss of status in the *realpolitik* of academic economics, referencing the recent availability of 'spatially oriented software', a 'paradigmatic shift' within their fields, and "the explosive growth of a global economy with its demands for location-based information".[7] Differences in editorial backgrounds and geographical location significantly inflect the nature of the two publications. Contributions to *Deep Maps and Spatial Narratives* feel to us literalist and instrumental rather than a combination of the scholarly and creative. This distinction marks a clear point of divergence within current deep mapping practice.[8]

Two-thirds of the articles in *Deep Mapping* discuss contributors' own creative projects, reflecting on their experience 'in the field' as practitioners of deep-mapping. These are either 'disciplinary' mappings, for example those by the archaeologist Carenza Lewis, or 'open' mappings, such as Laura Bissell and David Overend's "textual 'deep map' of . . . experimental commutes' made in the west of Scotland".[9] A direct engagement with mapping in its literal, cartographic sense is not a priority here, other than for the American geographer, Denis Wood, and there is a real sense of the openness of deep mapping practice in the concluding remarks of Roberts' preface:

> Whether or not we wish to call what emerges from this process a "map" (or the process itself "mapping") seems to me less important than the fact that it is taking place at all. In its most quotidian sense, then, deep mapping can be looked upon as an embodied and reflexive immersion in a life that is lived and performed spatially. A cartography of depth. A *diving within.*[10]

Deep Maps and Spatial Narratives (Spatial Humanities) is, by contrast, largely a collection of reflections on methodology aligned to digital innovation. With the notable exception of Stuart C. Aitken's *Quelling Imperious Urges: Deep Emotional Mappings and the Ethnopoetics of Space*, premised on the assumption that "the connections between technology and art, and places and power are by no means straightforward and progressive", its contributions are framed by familiar academic presuppositions that render all forms of deep-mapping work involving arts skills and practices as invisible.[11] 'Practice' here is, in an analysis of this version as it is applied, that of cartography reframed by GIS, computational models and similar manifestations that can be appropriated to a 'Digital Humanities'. In short, this book is largely framed by instrumental and analytic presuppositions that we wish to avoid here. These are reflected in the following statement which, accurate enough in its own terms, fails to acknowledge the poetics of paradox and ambiguity essential to open deep mapping as we understand it:

> A deep map is . . . an environment embedded with tools to bring data into an explicit and direct relationship with space and time; it is a way to engage evidence within its spatiotemporal context and to trace paths of discovery that lead to a spatial narrative and ultimately a spatial argument; and it is the way we make visual the results of our spatially contingent inquiry and argument.[12]

These different orientations reflect the Czech poet and immunologist Miroslav Holub's view that poetic activity, and by implication that of the arts more generally, is predicated on the *inadequacy* of its means.[13] It is precisely this inadequacy that enables the arts to evoke our lived experience as always exceeding and falling short of the conceptual definitions central to analytical thinking. Such inadequacy is, however, deeply problematic in the context of academic research, which insists on the necessity of generating adequate, or at least temporarily adequate, methodologies and categories (Figure 3.2).

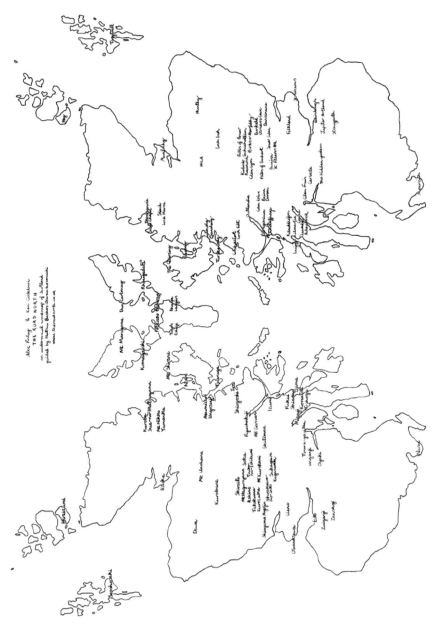

Figure 3.2   Alec Finlay, *The Road North*, word-map blog, book, audio, with Ken Cockburn, 2014.   Permission: Alec Finlay.

Roberts' *Deep Mapping* assumes a constellation of practices that are "multi-faceted, open-ended and – perhaps more contentiously – irreducible to formal and programmatic design".[14] His response to my own view that 'open' deep-mapping needs to avoid complicity in its own disciplining; to avoid the limiting by academic presuppositions of its capacity for informed, passionate and polyvocal engagement with a particular life-world, is telling in this respect. As he rightly points out, the need to work in a 'space in-between' both disciplinary and other voices highlights the need to tread "a fine line between complicity and creative dissolution".[15] In his view my "call for an 'open deep-mapping' only makes sense insofar as its openness is sufficiently diffuse as to do away with the very idea of deep mapping in the first place". The challenge then lies in acknowledging two contradictory tendencies; a desire to find a degree of coherence in the term while maintaining its necessarily loose, plural, non-proscriptive and open application. This is a challenge that might return us to Casey's differentiation, quoted earlier, between a position as "a posit of an established culture" and place, seen not as an object of academic discourse but as "an essay in experimental living within a changing culture".[16]

As a comparison of Silvia Loeffler's work and Gini Lee's projects (Chapter 8) indicates, such mapping will require those involved to invent appropriate processes in response to specific sites and on the basis of the various resources available to them, including the various models of deep-mapping to which they have access. For instrumental purposes this immersive process will no doubt continue to be referred to as 'landscape biography', or as 'deep', 'soft', 'mycelial', 'alternative' 'narrative', 'geo-poetic', 'fluid', 'wild' or 'counter-' mapping, or again as 'deep mapping/slow residency'. But, as those involved in these compound practices know, what they do is best only provisionally named or defined, purposely left open. To do otherwise is to risk their work being co-opted by limiting and limited forms of academic discourse masked by a reframing as 'Spatial Anthropology', 'Digital Humanities', 'Digital Archeology' 'Geography as Spatial Enquiry' and so on, with the distortions of colonisation that inevitably follow. While it is impossible to control reception, makers can and should acknowledge that the results of their practice will be judged by the terms used to frame it.

Why 'slow' residency?

One understanding of the term 'slow residency' shares the emphasis on temporal slowness introduced by Carlo Petrini's slow food movement (1983) and taken up by more general cultural shifts toward slowing down the pace of our engagements with core aspects of life.[17] It signals a conscious engagement with time and the temporal in our lives, including challenging the assumption that faster is necessarily better. However, this is only one level of the meaning of 'slowness', and should not be taken too literally. Ultimately our intention is to highlight a quality of attention (explored in Chapter 10) that requires working at the speed that allows us to attend as well and as appropriately as we can to the world.

To question how we use our time is to engage with some of our most basic values. Our use of time is in large part determined by our relationships with others. Some of this is simply a question of necessity; time spent on the work that sustains us as individuals or citizens, for example holding down a job or undertaking jury service. But time has also become a key focus for identity-oriented consumption, something signalled by the vast range of advertising that, directly or indirectly, tells us that this or that product will enable us to 'save time'. (The time 'saved' by our buying such devices must, of course, first to be earned by the time we spend earning the money to pay for them.) To ask how we 'spend' time is to examine our values as they present themselves in our daily life. This often leads to uncomfortable realisations about the limitations imposed on our choices by what the equation 'time = money' means in terms of our lived experience. However, as the example of Christine Baeumler's work (set out in Chapter 4) typifies, this difficulty has in turn prompted individuals to find ways of interweaving their concerns as neighbours or citizens with possibilities latent in their professional work, helping to create new amalgams of people and skill that can constitute communities of transverse action.

In this context 'slowness', slowing down, is primarily about taking time to consider, reflect, pay attention and work out what is an appropriate use of time. For example, Les Roberts' deep-mapping of 'island-ness' through a multi-media exploration of a motorway traffic island only needed a residency of one day and night. That was sufficient in terms of the qualities of attention required by the project. However, there is always a need to ask whether our notion of 'appropriateness' is instrumentally determined, or is a fully considered response to the requirements inherent in the qualities of the place itself. The Marises' considerations of slow residency and long-term commitment to the Loch Rannoch area of Scotland are discussed in Chapter 9.

As with so many aspects of deep mapping/slow residency, when thinking about slowness it is important to avoid being too literal. As Gini Lee writes of her *Oratunga* project, detailed in Chapter 8, in terms that point to the constraints imposed by academic governance, that it:

> cannot be entirely framed or realized according to prescribed timelines enforced by research programs. It is caught in a process of continual development and chance outcomes and endures as a work in progress.[18]

Such projects may produce 'outcomes' in the form of artworks, maps, publications, exhibitions, doctoral theses, and so on, which mark way stations in a process that their production may, or may not, conclude. Since we may return to the places that animates them, and in doing so reengage with them in ways that require new articulations, the 'conclusion' of such projects is often tacitly bracketed by the phrase 'for now'.[19]

But why 'residency', as we have used it here? The notion of 'residency' may appropriately refer to the length of time necessary to undertake an immersive engagement with a specific place, but this need not exhaust its meaning. In the arts and academic worlds 'residency' commonly refers to a period of time spent by an individual in another environment, something apart from home and the usual routine, so as to focus on a particular project or a resource not otherwise easily accessed. In this sense a residency provides time for reflection, research, discovery, new conversations and production of work. There is a sense, too, of protected time – a period of focus that is intense and determined. Residencies often have the possibility of allowing individuals to interact with a place or community other than that to which they are habituated, to take risks by engaging with new conditions and contexts. Of course, this deeper engagement can also be applied to the communities in which we find ourselves. All of which suggests an attitude of openness, concentration and immersion that is an important part of what we understand by 'slow residency' here.

Part Two: Deep mapping/slow residency

Some pre-histories of deep mapping?

Conventionally, deep-mapping is understood as emerging during the second half of the 20th century. Literally speaking, that is fair enough. However, phenomena may exist long before we recognise them for what they are. For example, old tensions and conflicts have the possibility to create new alternatives:

> Conflict from the viewpoint advanced here is not confrontation; it is conflict as engagement with the multidimensionality of human beings, their histories, social conditions, gender, languages and cultures.[20]

Similarly, mutations arise from existing circumstances, creating something new that didn't exist before. And even for those specific phenomena that do exist, there has not always been an awareness of these diverse and distinctive individual items. For example, there are names for only about 14% of all species currently living on the earth. This being so, we can suggest that the intuitions and energies that inform what we now call deep-mapping may have a tacit history going back at least to the 17th century and perhaps much earlier in intellectual history.

Giuliana Bruno argues that maps like the *Carte de Tendre* of Madeleine de Scudéry, later to be a major influence on the Situationists, constituted a new 'cartographic rendering of intimate experience'.[21] Counterpointed with the development of orthodox cartography, these opened up a tensioned 'space-between', given concepts of spatial organisation on one hand and how we are placed in the world on the other: a space-between that deep mapping/slow residency must always engage with in some way (see Figure 3.3). Those

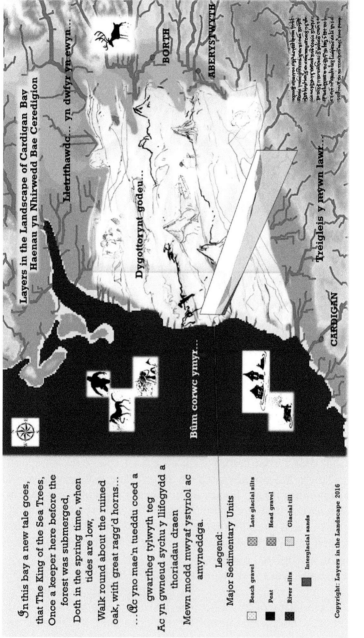

Figure 3.3 Erin Kavanagh, 'The Thin Deep Map' from *Layers in the Landscape* 2016, in *Rethinking the Conversation: A Geo-mythological Deep Map 2016*. Working as a geomythologist, artist and poet, Kavanagh combines both science and the arts in her deep mapping work. Photo: the artist.

17th-century intimate mappings took personal, emotional and bodily matters seriously, and, in doing so, constituted one aspect of an 'intimate revolt' that allowed composite narrative representations deploying cartographic devices to have a substantive impact on "the art of mapping today, particularly in artistic undertakings".[22]

Merlin Coverley identifies another tradition that informs both deep mapping and psychogeography. This was initiated by Daniel Defoe's *Journal of the Plague Year* (1722), a "blend of fiction and biography, of local history and personal reminiscence".[23] He links this to a British tradition of psychogeography, which includes the visionary London poet, William Blake. However, it is possible to trace this tradition back to Michael Drayton's *Poly-Oblion* (1598–1622), a vast and inclusive poetic sequence of 30 'songs' with accompanying maps in which each region of England and Wales is envisaged as a human figure. And this in turn resonates with compound mytho-cosmological mappings found in numerous non-Western cultures, such as the mapping of multiple worlds on the Shamanic drums that so inspired the artist, Wassily Kandinsky.

'American' deep mapping

These pre-histories are likely to seem somewhat eccentric in North America, where the terms 'deep map' and 'deep mapping' are usually associated either with an environmentally inflected regionalist form of documentary non-fiction, or with the instrumental application of technologies such as GPS. In the first case deep mapping is usually assumed to have originated with Wallace Stegner's *Wolf Willow: A History, a Story, and a Memory of the Last Plains Frontier*, although it could be argued that it begins with Henry David Thoreau's *Walden: Or, Life in the Woods*, a text central to Kenneth White's geopoetics. Stegner's text includes both fictional and nonfictional elements, historical material, impressions of the natural world, childhood remembrances and adult reflection on that childhood. As such, his situated yet polyvocal approach anticipates William Least Heat-Moon's *PrairyErth (a deep map)*, from which the term 'deep mapping' derives.

In 1991 William Least Heat-Moon (born in 1939 and christened William Lewis Trogdon) published *PrairyErth (a deep map)*.[24] This is an extensive exploration of Chase County, Kansas, as the last remaining expanse of native tall-grass prairie in the USA. In terms of deep mapping as a literary 'method', it can be read as both developing and adjusting Stegner's approach, enlarging on its range of references and environmental concerns in the context of 'participatory history'. However, there is another, equally significant, aspect of this book. It avoids conservative Regionalist tropes and engages in a 'formal playfulness' and breadth of reference that mimics Laurence Sterne's *The Life and Opinions of Tristram Shandy, Gentleman* (1769). In setting up creative tensions and multiple perspectives that work between regionalism, environmentalism and modernist tendencies, it resonates with what the architectural writer and theorist, Kenneth Frampton, calls Critical Regionalism (1988, 1983). These generate *Prairy Erth's* wide-ranging environmental mediation, one able to combine something of the sense of enchantment that Jane Bennett advocates, with a sense of eco-logical loss and absence.[25] This in turn allows it to avoid the pull towards nostalgia and guilt that permeates *Wolf Willow*. The term deep-mapping, derived from *PrairyErth (a deep map)*, appeared soon after its publication, with Michael Shanks claiming that he and Mike Pearson 'invented' the con-cept of the deep map in 1994 '(after William Least Heat-Moon)'.[26]

The North American view of deep-mapping as a literary genre has tended to marginalise other, related approaches with the consequence that, until re-cently, they attracted little academic attention.[27] Nor has any attempt been made to consider the relationship between the Regionalist genre of deep mapping and the 'counter-mapping' identified in 1995 by Nancy Peluso. A political ecologist, Peluso uses ethnographic and historical material to study the social processes affecting management of land-based resources. Her view of counter-mapping as contesting the power of the self-serving state or corporate cartographic accounts thus begs the question as to the underplaying of the political, implicit or explicit, in reflection on deep map-ping as 'vertical' or 'deep' travel writing. Europe, by contrast, has produced a diffuse body of writing concerned with detailed narrative mappings of place, including John Copper Powys' *Wolf Solent* (1929) and *A Glastonbury Romance* (1933), Tim Robinson's work, most notably his two-volume *Stones of Aran* (1985) and, perhaps more peripherally, Peter Davidson's *The Idea of the North* (2005). This suggests that a literary urge to deep-mapping is broader and more international than is sometimes acknowledged.

We might ask whether Ursula Le Guin's *Always Coming Home*, a unique fictional work first published in 1985, is a form of deep mapping? A highly creative collaboration between the novelist and artist Margaret Chados-Irvine, composer Todd Barton and Geomancer George Hersh, it offers an ecologically informed mapping of 'the Valley', situated in California. Framed by an introductory section entitled 'Towards an Archaeology of the Future', it can be read as a literary variation on Pearson and Shanks' argument for deep-mapping as "a *blurred genre* . . . or a science/fiction, a mixture of narration and scientific practices, an integrated approach to re-cording, writing and illustrating the material past", using the conventions of science fiction.[28] Similarities between the social concerns of Le Guin's archaeological framing and the theatre/archaeology of Pearson, Shanks and Cliff McLucas are apparent in her account of 'the gap between'.[29] It is also evident in her articulation of a situated culture in which:

> no distinction is made between human and natural history or between objective and fact and perception, in which neither chronology nor causal sequence is considered an adequate reflection of reality, and in which time and space are so muddled together that one is never sure whether they are talking about an era or an area.[30]

Furthermore, Le Guin's attention to the narratives, epistemology and cosmology of native Americans in California resonates with Rebecca Solnit and *PrairyErth*'s challenge to the Edenic presuppositions about creation that underpin assumptions shared by much mainstream Euro-American environmental thinking, along with those of artworks framed by that perspective. Solnit, however, contrasts these assumptions by referencing 'Native American creation stories' that present "a worldview in which creation of the world is often continual and sometimes comic improvisation, without initial perfection or a subsequent fall".[31]

Psychogeography and deep-mapping: Europe and elsewhere

The various European traditions relevant to deep mapping/slow residency are usually framed in terms of site-specific performance art on one hand (Kaye 2000), and the tradition of literary psychogeography derived from the Lettrist International in the 1950s on the other. Psychogeography can be seen as the end product of a tortuous, largely French, tradition that derives from the 17th-century 'intimate mappings' tradition, Charles Baudelaire's *flâneur* (or *flâneuse*, to recognise a female form of experience), and Walter Benjamin's *Arcades Project*, running through the work of both Surrealists like André Breton, and the anti-Surrealists, or para-Surrealists René Daumal and Roger Gilbert-Lecomte.

The term 'psychogeography' itself is derived from the activities of the Lettrist International (1952–57) and the Situationist International (1957–72) and, more specifically, from the writing of Guy Debord. Those working in English identified with this tradition include the writers Peter Ackroyd, J. G. Ballard and Iain Sinclair, the journalist and novelist Will Self, the political polemicist and prankster Stewart Home, and the radical artist/filmmaker Patrick Keiller. This tradition foregrounds walking practices (largely urban) and is constantly being revised and re-evaluated, most recently by Lauren Elkin's *Flaneuse: Women Walk the City in Paris, New York, Tokyo, Venice and London* (2016) and Phil Smith's *The Crisis in Psychogeographical Walking: From Paranoia to Diversity, Ecology and Salvage*.[32] However, although perceived as predominantly a literary genre, it is also associated with significant performance work, for example by the British performance collective *Wrights & Sites*.[33] More of these contemporary and emergent variations on walking practices will be touched upon in Chapter 7, particularly with the research of Gareth M. Jones.

The film-maker Patrick Keiller's work can be seen as bridging the largely urban pre-occupations of psychogeography and the more often rural or post-rural sites of deep mappings. Keiller studied architecture and environmental media and has taught both architecture and fine art students at various times. He made his first film in 1981 and is best known for the films *London* (1994), *Robinson in Space* (1997) and *Robinson in Ruins* (2010), all of which offer a particular filmic inflection of the politics of psychogeography, without referring exclusively to urban environments.[34] The constituency and operation of the collaborative team (which included the political geographer Doreen Massey), with whom he researched and made *Robinson in Ruins*, together with the 2012 installation *The Robinson Institute* at Tate Britain is indicative of one form of collaborative deep mapping. A highly creative and imaginatively nuanced exposition of the relationship between politics, history and place, *The Robinson Institute* included film, curated artefacts, and carefully researched polyvocal texts that interwove diverse material in telling juxtapositions. Perhaps the recent work nearest in spirit and ambition to *The Robinson Institute* has been the *City, Kaavad and Other Works* exhibition of the painter Gulammohammed Sheikh, held at the Vadehra Art Gallery of the Lalit Kala Academy in New Delhi in 2011.[35]

Something of why this is the case can be indicated by noting that the centre-piece of the exhibition was based on a *kaavad*, a mobile shrine containing painted narratives used by travelling performers as an aid to storytelling, with its doors opened or closed in different sequences to combine multiple images in ways that anticipate a final revelation. Sheikh used this format to create combine various permutations of images and narratives that provide various ways of examining past and present social, cultural, political belief systems. While the form of the exhibition owes nothing to deep mapping, its richness of concern and reference and its compound narrative spirit is similar to that of *The Robinson Institute*.

There are a range of deep-mapping/slow residency practices and projects, for example in China and Australia, that resonate with our concerns in this book. One example would be the Iranian artist and architect Tahmineh Hooshyar Emami's use of the spatial elements in *Alice in Wonderland* and *Alice through the Looking Glass* to map the experience of displacement experienced by refugee children in Europe. Another would be Perdita Phillips' *The Sixth Shore* (2009–14), a research project that created an audio soundscape or 'layered map' that engages the listener with a physical and social environment of Lake Clifton, near Mandurah, Western Australia, that is layered in both space and time. This flows from applying ecosystem thinking to the different spatial and temporal scales of a specific location, prompted by a reef of microbial thrombolites in the lake with an ancestry that can be traced back to fossils three and a half billion years old.[36] The work draws attention to both the deep and recent history and ecology of the lake, which was only separated from the sea 4,000 years ago.[37]

Out of Wales: Theatre/Archaeology and performative deep-mappings

In Europe, and in performance and archaeological circles more generally, the term 'deep mapping' is largely associated with a site-based performance practice, also known as 'theatre/archaeology' or 'performance archaeology' that originates in work by Mike Pearson, Michael Shanks and Clifford McLucas with the radical Welsh performance group *Brith Gof*. After

McLucas' untimely death in 2002, Pearson and Shanks took this impulse in different, but not unrelated, directions. Pearson continued to develop site-based performances that work "horizontally across the terrain and simultaneously vertically through time" to become "a topographical phenomenon of both natural history and local history", while Michael Shanks engaged with an expanded notion of archaeology.[38] Their introduction of the term deep mapping has subsequently been adopted by a variety of individuals working in visual and performative arts, digital arts, architectural education (notably by Sophia Meeres in the collective 'landscape biography' of Arklow), and landscape curation to describe their own landscape or place-oriented work and concerns.

Historically, the notion of performance/archaeology appears to have emerged from Mike Pearson's involvement with the Cardiff Laboratory Theatre and the 'theatre anthropology' of Eugenio Barba's *Odin Teatret*, which toured Wales in 1980. In 1981 Pearson, a former archaeology student, co-founded *Brith Gof* with Lis Hughes Jones, and became its first artistic director. *Brith Gof* is now internationally acknowledged as a pioneering experimental performance group dealing with place, identity and the role of the presence of the past in strategies of cultural resistance and community construction. Clifford McLucas joined the company in 1988, shifting its work in a more explicitly site-specific direction.

Shanks and Pearson's reading of *PrairyErth* in the early 1990s led to an extended understanding of both their training as archaeologists and of the particularities of place articulated by Welsh terms such as *yr aelwyd* (the hearth), *y filltir sqwar* (the square mile), *yo fro* (neighbourhood, home district, *heimat*) and *cynefin* (habitat).[39] Subsequently *Brith Gof* produced powerful and flexible multi-media performances on a range of scales that engaged with "the matrix of particular folds and creases, the vernacular detail, which attaches us to place"; a practice that, in retrospect, can be seen to both parallel, and extend through particularisation, disciplinary based expositions of place from cultural, philosophical and geographical perspectives.[40]

Pearson, now an Emeritus Professor of Performance Studies at Aberystwyth University, went on to design one of Britain's first undergraduate degrees in Performance Studies in the late 1990s and, as a performance maker, teacher and researcher, he has remained a highly influential figure in both the national and international fields of theatre and performance. An influential example of his more recent performance work is documented in his *'In Comes I': Performance, Memory and Landscape* (2006), a work that shares certain points of contact with both the literary psychogeography of Iain Sinclair, Peter Ackroyd and W. G. Sebald, and with the combination of visual and narrative mappings of an individual like Luci Gorell Barnes, particularly her project *The Stinging Nettle Atlas*, a project concerned, like Pearson's, with the 'placings' of childhood.

Michael Shanks, now the Omar and Althea Hoskins Professor of Classical Archaeology at Stanford University, is perhaps best known for work with Christopher Tilley on post-processual or interpretive archaeology. Given its eclecticism at both the theoretical and technical levels, along with Shanks' support for the continuing relevance of Cliff McLucas' work, his approach remains closely linked to that of deep mapping. However, it is predicated on, and intended to privilege, the discipline of archaeology conceptualised 'as a mode of cultural production' or 'cultural poetics'.[41] What we find problematic about this disciplinary appropriation appears in his claim that 'we are all archaeologists now', alongside his view of the Arts and Humanities as a 'fascinating research laboratory'. These claims reflect a tacit intellectual neo-colonialism typical of the combative *realpolitik* and disciplinary empire-building endemic to the academy. (Similar attempts at appropriation can be found in, for example, the writing of certain cultural geographers). Something which, in turn, imposes arbitrary limits on open deep-mapping/slow residency as a knowledgeable, passionate, multi-perspectival and open engagement with the world as polyverse.

Cliff McLucas: deep mapping the island of Terschelling

While Cliff McLucas' deep mapping work in California is known from on-line archives, his final deep mapping project is largely forgotten. (What information we have comes from a performed lecture and from Joop Mulder, McLucas' collaborator and the director of the *Oerol* festival.)[42] However, it suggests that his work was developing rapidly and in new directions. Between 1999 and 2002 McLucas initiated a major new project. Its first iteration, *Earste Dagen (The First Days)*, was presented at the *Oerol* annual festival on the island of Skylge/Terschelling, located off the Netherlands mainland in the Wadden Sea. *Earste Dagen* was intended as a pilot project, undertaken with the view of developing methods for deep-mapping places and peoples by combining cartography, aerial surveys and other 'neutral' technological processes with recordings reflecting the political and social concerns (in particular with regard to language) relating to the island's history and inhabitants. From what we know of what McLucas presented in 2000, it appears that he was preoccupied with the fragility of the island (geologically little more than a large, artificially stabilised, sandbank) that even slight changes in sea level would submerge.

Working with photographer Fred Ross, who used a helicopter to take the necessary photographs, the installation presented the island as wholly isolated; with McLucas combining 'an aerial video survey of the island, a wild track sound recording from five different locations on the island, an interview with the oldest man on the island [and a *Fris* speaker], and a series of digital stills'.[43] These were installed in a Second World War bunker, a former canteen of the German Wehrmacht, which further emphasised his concern with fragility. It seems that the central focus of the work was a projection onto the reinforced concrete floor, involving a large map of Terschelling. While this does not imply any radical departure from *There are ten things that I can say about these deep maps*, sumptuousness appears to have given way to a materially appropriate response to the fragility of both the land and its original language of Frisian or *Fries*.[44]

McLucas' preoccupation with language in his deep mapping derives from his learning, and concern for the future of, the Welsh language. Asked about the Terschelling project's focus on language, he said:

> In fifty or a hundred years' time . . . maybe the language will have changed, for sure it would have changed I think, and that's why it's exciting for us to make a portrait of this point in time.[45]

Anwen Jones and Rowan O'Neill rightly reference McLucas' concern with the British perspective on Welsh culture as 'underground' or 'private' and his stress on deep-mapping's potential to address the place that such a "culture inhabits – both geographical and emotional (dare I say 'spiritual'?)".[46] This concern is present in the Californian work in his insistence on including the languages of Native Americans spoken along the length of the San Andreas Fault. It also resonates with the work of many contemporary practitioners who embrace this importance of language; for example in Tim Robinson's work on the west of Ireland, artist Arthur Watson's work in the inclusion of Scots place names, and Dakota artist Mona Smith's *B'dote* memory mapping, all concerned to address the fragility of the intimate relationship between language, local naming and a sense of place. That McLucas recognised the importance of such concerns and had begun to link these to environmental issues remains significant for current approaches to deep mapping/slow residency.

The disciplinary environment/the authority of open practice

Silvia Loeffler, whose *Glas Journal* was referred to earlier, has an ensemble practice of the kind referred to in the Introduction. However, this term needs to be thought through in relation to how such practices are authorised. She refers to her chronotopic approach to her residency as a form of deep-mapping, but also as a 'hybrid ethnographic project' concerned with 'the cultural mapping of spaces we intimately inhabit', using the terms of 'liquid-' and 'tender-' mappings that refer to Giuliana Bruno's discussion of Madeleine de Scudéry's *Carte du pays de Tendre*.[47] This compound naming seems entirely appropriate for a collaborative, many-faceted project that explores the layered emotional geographies of Dún Laoghaire Harbour, Dublin, by focussing on "performatively mapping the intimate rituals and everyday performances of those individuals who live and work in the harbour" (Figure 3.3).[48] She adds that by developing the project with the participation of local inhabitants, she has been able to explore the maritime environment as a liminal (or, arguably, 'undisciplined') space, one in which the character of buildings, the area's economic implications, daily spatial rhythms and a sense of safety all help determine relationships to space. The project is, then, as ambitious as it is complex. Loeffler engaged with 14 individuals who live and work in the harbour area to produce handmade books that will constitute a record of what the harbour space means to the residents based in the old coastguard station, along with individuals involved with a host of other harbour related organisations and clubs. This suggests that what is particularly valuable about such projects is precisely its richness and complexity of reference and evocation, which she refers to in terms of the interactions between persons and their habitat, and as existing in the constant flux of appearance, disappearance and reappearance, one which might be seen as analogous to a tidal system that regulates liquid states, times and places.

At a time when cultural geography and other academic disciplines in the social sciences are seeking to expand their fields of influence and expertise by establishing "new ways of activating some form of 'productive nexus' with art that 'moves beyond familiar, discursive, models of interdisciplinarity to engage seriously with the immediate material efficacies of contemporary art as a mode of spatial enquiry'", it is important to ask how this relates to deep mapping and slow residencies.[49] In short, where does the authority of deep mapping/slow residence as a process by which individuals are able to articulate such 'a constant flux of appearance, disappearance and reappearance' reside? As creative individuals drawing on a broad range of arts skills, initiators of deep-mapping/slow residency projects strive to create work that 'speaks for itself'. However, their likely entanglement in the world of academic research frequently requires such individuals to engage in detailed textual expositions of their work as a way of 'justifying' their own position as authoritative.

Such 'justification' requires, in turn, that the richness and complexity of reference in the work itself is matched in such accounts, particularly in terms of what the author regards as authoritative. In this respect *Glas Journal: Deep Mappings of a Harbour or the Charting of Fragments, Traces and Possibilities* is an exemplary model of exposition. It avoids the presuppositions of a disciplinary framing and references the work of Mikhail Bakhtin, anthropological and ethnographic texts, a range of visual artists' work, art criticism, work by social, political and cultural geographers, literature, cultural history, film, poetry, linguistics (including the etymology of Homer's 'wine-dark sea'), local history, landscape studies and material on deep-mapping. This breadth of reference to what Loeffler regards as 'authoritative' is in marked contrast to normative disciplinary textual practice, where authors largely cite material from recent contributions to their own specialist disciplinary field.

In making this point we aim to remind both ourselves and our readers that reference to philosophers and theorists currently adopted by disciplinary authorities – for example, Bruno Latour, Gilles Deleuze, Isabelle Stengers, Felix Guattari, or indeed theorists associated with speculative realism (all of which work we value, and indeed have cited ourselves within the pages of this book) – as authoritative is no substitute for an informed and open creative engagement with dialogic, communicative thinking in the life-world. Arguably, a rule-of-thumb measure of the seriousness of any disciplinary scholar who seeks 'a productive nexus with art as a mode of spatial enquiry', is the degree to which her or his published texts reference specific examples of the creative arts as authoritative in terms other than those of their own discipline. In other words, the creative arts are primary texts for understanding the world.

The importance of this issue of the 'horizontality' of claims to authority with regard to open deep-mapping/slow residency, can be read through a discussion of Denis Wood's account of what he refers to as "an *avant la lettre* deep mapping project", one he carried out as a geographer undertaking studio-based teaching with landscape architecture students at the School of Design at North Carolina State University in the early 1980s.[50] This work, carried out through neighbourhood mapping projects, eventually led to the production of the copious material published as *Everything Sings: Maps for a Narrative Atlas*.[51] His description of the neighbourhood "as a transformer, turning the stuff of the world (gas, water, electricity) into the stuff of individual lives (sidewalk graffiti, wind chimes, barking dogs), and vice versa" within the context of undergraduate education might seem mundane.[52] However, nothing could be further from the truth.

Denis Wood's passion for genuinely educating students who knew nothing about cartography is clear when he asks them "to map the way the land smelled, the way it felt in their legs when they walked it, the way twilight made all the difference."[53] That Wood was attuned to the openness of deep-mapping well before he knew the term is illustrated by his account of making an 'inefficient map' of the experience of street lights, which he describes as having: "a sense of poetry, something imagistic, a little like Pound's 'The apparition of these faces in the crowd;/Petals on a wet, black bough' . . . [that] might manifest on a map, a map attentive to the experience of place"; an experience that led him to "thinking seriously about a poetics of cartography".[54] His passion for what was liberating in the arts of modernity (in addition to Pound he references Schoenberg, Arp, William Carlos Williams, Lawrence Durrell's *Clea*, and *Singing in the Rain*) is balanced by a very clear sense of the ways in which the implicit politics of cartography as a specialist discipline impact on the distribution of power, on voting patterns, wealth, education and health, as well as the outdated empiricism that underwrites cartography's commitment to "the failed rationalities . . . the empty harmonies . . . the make-believe coherences of Enlightenment, of Victorian thinking".[55] In what might be heard as a repost to the editorial approach of *Deep Maps and Spatial Narratives (Spatial Humanities)*, and to the disciplinary academy's often uncritical adoption of new technology, Wood concludes by stressing the importance of mapping, not as "dropping some data into a computer mapping program" but as "getting out and doing the fieldwork, of

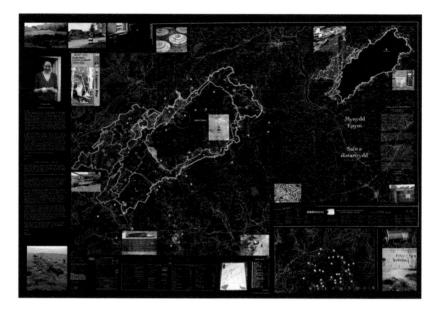

Figure 3.4 Iain Biggs *Hidden War (with and for Anna Biggs)* from Iain Biggs
"'Listening at the Borders": introduction, acknowledgements (and an in-
tervention)' in Iain Biggs, ed., *Debatable Lands Vol. 2. These Debatable
Lands* (Bristol: Wild Conversations Press, 2009). Photo: Iain Biggs.

"looking hard at stuff". . . and of engaging with place directly, on foot, because doing that and recording the results generates "an irreplaceable kind
of attention".[56] We could not agree more (see Figure 3.4).

Narrative and dialogic mappings

Luci Gorell Barnes refers to aspects of her diverse 'slow' projects – concerned
primarily with connection and belonging, she writes, makes paintings,
books, maps and animations – as narrative mappings. These include *The
Atlas of Human Kindness* (see Figure 3.5), which she describes as concerned
with community cartographies of compassion. This comprises a growing
collection of maps made by individuals and groups in Bristol, including
refugee groups and children with learning difficulties. It shows where and
when they experienced kindness from people concerned about their rights,
feelings and welfare. It invites debate about how stories, memories and im-
aginings make and re-make place, and how fragmented personal landscapes
can become less fragmented, by inviting people to think individually and
collectively about values and connections, and about what networks and
community means. Gorrell Barnes adopts various strategies in making this
type of work. For example, she writes:

Figure 3.5 Luci Gorell Barnes and individuals contributing to *The Atlas of Human Kindness*. Photo: Luci Gorell Barnes.

I erected a workshop tent in the street outside my house, taped a large sheet of blank paper to a table top, and invited people to contribute stories of when and where they received kindness from another person.[57]

Reconfiguring the impulse to place-making

This is not a book solely about deep-mapping, but about finding appropriately creative ecosophical responses to the complex and often bewildering questions and concerns that appear when we engage with place. Deep-mapping happens to provide a rich cluster of interwoven practices that we can draw on when faced with those questions and concerns. We need such practices because, as Edward S. Casey's definition quoted earlier reminds us, a place is dynamic, always open and changing. Yet as the geographer Tim Cresswell notes, society tends to connect geographical place to assumptions about normative behaviour. Our language is shot through with assumptions that link social hierarchy to physical location. "Someone can be 'put in her place' or is supposed to 'know his place'. There is, we are told, 'a place for everything and everything in its place'".[58] This tension increases as industrial globalisation accelerates environmental issues like climate change, and starts to impact on communities.

As an artist, environmental activist and educator, Simon Read has learned over many years that our eco-social problems require "a particular kind of strategy that our culture has yet to develop and promote", the new way of thinking and acting we are advocating here.[59] This should not be confused with the increasing institutional emphasis on 'resilience', on meeting environmental change by building up 'social capital', 'community identity', 'place attachment', 'community cohesion' or 'community participation', concepts that are all-too-often imposed by officialdom, usually without much reference to the debates within communities or to their lived experience. As an alternative to this instrumental approach to resilience, we need to remember that human beings have the ability to take risks with who they are. They can choose, for example, to face unknown situations that they know will irrevocably change them. Our ability to risk everything differentiates us from natural systems. In ecological science, 'resilience' refers to an eco-system's ability to fend off or manage threats that would undermine its core characteristics.[60] The United Nations defines resilience as: "the capacity of a system, community or society potentially exposed to hazard, to adapt by resisting or changing in order to reach and maintain an acceptable level of functioning and structure".[61] However, as Brad Evan and Julian Reid point out, the politics of resilience focusses on adapting the status quo, *not changing it*. We, however, need to choose to risk genuine change, with a view precisely to moving away from the characteristics that we have allowed to determine us, our society and our environment as defined by the toxic culture of possessive individualism. So rather than use the conservative and often misleading term 'resilience', it makes more sense to think in terms of developing new strategies that help us act differently.

In his Preface to *Deep Mapping*, Les Roberts identifies the paradox that haunts this chapter when he observes that insistence on an open form of deep-mapping is paradoxical; it only makes sense insofar as that openness is sufficiently diffuse as to do away with *the very idea of deep-mapping as a given category*. But this claim sees deep mapping from an academic perspective. Any ensemble self must live with the realities of both the dynamics of place and the realities of practice, understanding their inevitable instability. As a condition of 'being-as-becoming'. This view is beautifully reflected in Elizabeth-Jane Burnett's *The Grassling*, a poetic homage to her father, their familial place, and to language, that can be read as an extension of deep mapping into the realm of eco-poetics.

Writing about *Either Side of Delphy Bridge*, a deep-mapping undertaken in north Cornwall, Jane Bailey and I describe our working process as made up of "observing, listening, walking, conversing, writing and exchanging . . . of selecting, reflecting, naming and generating . . . and of digitalizing, interweaving, offering and inviting".[62] Taking up Lee Roberts' reflections on that claim, it seems that we were no more conducting 'a mode of spatial enquiry', or indeed any other activity as academically categorised, than we were making a deep map. First and foremost, as Roberts himself notes, we were immersing ourselves "in the warp and weft of a lived and fundamentally intersubjective . . . creative coalescence of structures, forms, affects, energies, narratives, connections, memories, imaginaries, mythologies, voices, identities, temporalities, images and textualities".[63]

Finally, perhaps all that can usefully be said is that such work is an embodied, open and reflexive immersion in a particular grounded polyverse, one that is as much temporal as spatial, political as poetic, testimonial as speculative, imaginal as analytic. For instrumental purposes this immersive process may be called 'landscape biography', or 'deep', 'soft', 'mycelial', 'alternative' 'narrative', 'geo-poetic', 'fluid', or 'counter' mapping. But those directly involved in it know perfectly well that what we are doing is best left open, free of fixed definition, as affirmed recently by Erin Kavanagh in her discussion of its role in mediating between the presuppositions of art and science.[64] This mediation, and our role in it, will be considered in the chapters that follow.

Notes

1 From Cliff McLucas, "I Was Invited to This Island", quoted in Anwen Jones and Rowan O'Neill, 'Living Maps of Wales: Cartography as Inclusive, Cultural Practice in the Works of Owen Rhoscomyl (Arthur Owen Vaughan) and Cliff McLucas', *International Journal of Welsh Writing in English*, 2, no. 1 (October 2014), 120.
2 Jones and O'Neill, 'Living Maps', 150.
3 Silvia Loeffler, '*Glas Journal*: Deep Mappings of a Harbour or the Charting of Fragments, Traces and Possibilities', in *Deep Mapping*, ed. by Les Roberts (Basel: MDPI AG – Multidisciplinary Digital Publishing Institute, 2016), 30–48. Silvia Loeffler, 'Glas Journal 2015', <https://silvialoeffler.wordpress.com/about/transit-gateway-4-frenzy-and-excitement/transit-gateway-3-structures-of-care/transit-gateway-2-walls-of-protection/transit-gateway-1-a-shoreline-of-anxiety/glas-journal-2016/glas-journal-2015/> [accessed 10 April 2020].
4 Barny Warf, 'Deep Mapping and Neogeography', 139; Ian Gregory and Others, 'Spatializing and Analyzing Digital Texts: Corpora, GIS and Places'; Martin Worthy, 'Warp and Weft on the Loom of Lat/Long', 217: all in *Deep Maps and Spatial Narratives*, ed. by David Bodenhamer, John Corrigan, and Trevor Harris (Bloomington: Indiana University Press, 2015).
5 Owain Jones, 'Stepping from the Wreckage: Geography, Pragmatism and Anti-representational Theory', *Geoforum*, 39 (2008), 1600.
6 *Deep Mapping*, ed. by Les Roberts (Basel: MDPI AG – Multidisciplinary Digital Publishing Istitute, 2016); David Bodenhamer, John Corrigan, and Trevor Harris, eds, *Deep Maps and Spatial Narratives* (Bloomington: Indiana University Press, 2015).
7 Bodenhamer, Corrigan and Harris, *Deep Maps*, 2.
8 However, there is obviously a degree of overlap between these divergent positions. The prospectus for Spatial Humanities workshops run by the American, Lincoln Mullen, a digital historian working on the history of American religions, makes it very clear that it is not simply questions of geography that account for different evaluations of ambiguity, layered-ness, multiplicity of data, perspectives and voices, or narrative engagement with issues such as power and identity. See: Lincoln Mullen, 'Deep Maps', Spatial Humanities Workshop

(2015) <http://lincolnmullen.com/projects/spatial-workshop/deep-maps.html> [accessed 10 April 2020].

9 For a discussion of 'disciplinary' and 'open' deep mapping, see: Iain Biggs, 'The Spaces of "Deep Mapping": A Partial Account', *Journal of Arts and Communities*, 2, no. 1 (2011), 5–25. For information about Bissell and Overend's work see: Laura Bissell and David Overend, 'Regular Routes: Deep Mapping a Performative Counterpractice for the Daily Commute', *Humanities: Deep Mapping*, ed. by Les Roberts (Basel: MDPI AG - Multidisciplinary Digital Publishing Institute, 2016), 131.

10 Les Roberts, 'Preface: Deep Mapping and Spatial Anthropology', in *Deep Mapping*, ed. by Les Roberts (Basel: MDPI AG - Multidisciplinary Digital Publishing Institute, 2016), ivx.

11 Les Roberts, 'The Rhythm of Non-Places: Marooning the Non-Self in Depthless Space', in *Deep Mapping*, ed. by Les Roberts (Basel: MDPI AG – Multidisciplinary Digital Publishing Institute, 2016), 103.

12 Bodenhamer, Corrigan and Harris, *Deep Maps*, 3–4.

13 Miroslav Holub, *The Dimensions of the Present and Other Essays*, ed. by David Young (London: Faber & Faber, 1990), 145.

14 Roberts, *Deep Mapping*, ivx.

15 Ibid., xi.

16 Edward S. Casey, *Getting Back into Place: Toward a Renewed Understanding of the Place-world* (Bloomington: Indiana University Press, 1993), 31.

17 Carlo Petrini, *Slow Food: The Case for Taste* (New York: Columbia University Press, 2001).

18 Gini Lee, 'The Intention to Notice: The Collection, the Tour and Ordinary Landscapes' (Unpublished Doctoral dissertation, RMIT University, 2006).

19 Lee, 'Intention to Notice'.

20 Michael Cronin, 'Who Fears to Speak in the New Europe? Plurilingualism and Alterity', *European Journal of Cultural Studies*, 15, no. 2 (2012), 190.

21 Giuliana Bruno, *Atlas of Emotion: Journeys in Art, Architecture, and Film* (London: Verso, 2002), 24.

22 Bruno, *Atlas of Emotion*, 235.

23 Merlin Coverley, *Psychogeography* (Harpenden: Oldcastle Books, 2006), 36–37.

24 William Least Heat-Moon, *PrairyErth: A Deep Map* (Boston, MA: Houghton Mifflin, 1991).

25 Jane Bennett, *Vibrant Matter: A Political Ecology of Things* (Durham, NC: Duke University Press, 2010).

26 <http://traumwerk.stanford.edu: 3455/MichaelShanks/886> [accessed 19 January 2008].

27 As exceptions to this, see Cathy Turner, 'Palimpsest or Potential Space? Finding a Vocabulary for Site-Specific Performance', *New Theatre Quarterly*, 20, no. 4 (2004), 373–390; Nick Kaye, *Site-Specific Art: Performance, Place and Documentation* (Milton Park: Routledge, 2000).

28 Mike Pearson and Michael Shanks, *Theatre/Archaeology: Disciplinary Dialogues* (London: Routledge, 2001), 131.

29 Ursula K. Le Guin, Todd Barton, and Margaret Chodos-Irvine, *Always Coming Home* (Berkeley: University of California Press, 1986), 74.

30 LeGuin, Barton and Chodos-Irvine, *Always Coming Home*, 153.

31 Rebecca Solnit, *Wanderlust: A History of Walking* (New York: Penguin, 2001), 12.

32 Phil Smith, 'The Crisis in Psychogeographical Walking: From Paranoia to Diversity, Ecology and Salvage', in *Walking, Landscape and Environment*, ed. by David Borthwick, Pippa Marland and Anna Stenning (London: Routledge, 2020), 187–202.

33 Wrights & Sites, see: <http://www.mis-guide.com> [accessed 10 April 2020].
34 *Robinson in Ruins* was researched and made over a three-year period with financial support from the British Arts and Humanities Research Council's Landscape and Environment Programme. It is a collaboration between Keiller; the
   political geographer Doreen Massey, Patrick Wright (a professor of literature
   and visual and material cultures), and was accompanied by a related doctoral
   project – Matthew Flintham's, *Parallel Landscapes: A Spatial and Critical Study
   of Militarised Sites in the United Kingdom* (unpublished Doctoral dissertation,
   London: Royal College of Art, 2011).
35 For an insightful discussion of this exhibition see Karin Zitzewitz, 'Past Futures
   of Old Media: Gulammohammed Sheikh's Kaavad: Travelling Shrine: Home', in
   Media and Utopia: History, Imagination, Technology, ed. by Arvind Rajagopal
   and Anupama Rao (New Delhi: Routledge, 2016 pp. 189–209). A fully illustrated
   version was published on CSSAAME Borderlines, 11 March 2016.
36 Thrombolites are prehistoric clumps of microbial organisms that form mounds
   and photosynthesise.
37 See: Perdita Phillips, the sixth shore 2009–2014 (n.d.) <http://www.perditaphil-
   lips.com/current-projects/the-sixth-shore/> [accessed 10 April 2020].
38 Michael Pearson, *'In Comes I': Performance, Memory and Landscape* (Exeter:
   University of Exeter Press, 2006), 3.
39 Pearson, *'In Comes I'*, 14.
40 Pearson and Shanks, *Theatre/Archaeology*, 138–139. See: Lucy Lippard, *The Lure
   of the Local* (New York: The New Press 1997); Edward S. Casey, *Getting Back
   into Place: Towards a Renewed Understanding of the Place-World* (Bloomington:
   Indiana University Press, 1993); and Doreen Massey, 'A Global Sense of Place',
   Marxism Today, 38 (1991), 24–29.
41 Pearson and Shanks, *Theatre/Archaeology*, 50.
42 Cliff McLucas, 'I was invited to this island', performed lectures, available in
   the Cliff McLucas Collection, MCLT, National Library of Wales, Aberystwyth,
   <http://www.archaeographer.com/keyword/cliff%20mclucas/>.
43 Jones and O' Neill, 'Living MAPS', 116.
44 Cliff McLucas:

> *There are ten things that I can say about these deep maps ...* **First** Deep maps
> will be big – the issue of resolution and detail is addressed by size. **Second**
> Deep maps will be slow – they will naturally move at a speed of landform or
> weather. **Third** Deep maps will be sumptuous – they will embrace a range of
> different media or registers in a sophisticated and multilayered orchestra-
> tion. **Fourth** Deep maps will only be achieved by the articulation of a variety
> of media – they will be genuinely multimedia, not as an aesthetic gesture or
> affectation, but as a practical necessity. **Fifth** Deep maps will have at least
> three basic elements – a graphic work (large, horizontal or vertical), a time-
> based media component (film, video, performance), and a database or archi-
> val system that remains open and unfinished. **Sixth** Deep maps will require
> the engagement of both the insider and outsider. **Seventh** Deep maps will
> bring together the amateur and the professional, the artist and the scientist,
> the official and the unofficial, the national and the local. **Eighth** Deep maps
> might only be possible and perhaps imaginable now – the digital processes
> at the heart of most modern media practices are allowing, for the first time,
> the easy combination of different orders of material – a new creative space.
> **Ninth** Deep maps will not seek the authority and objectivity of conventional
> cartography. They will be politicized, passionate, and partisan. They will
> involve negotiation and contestation over who and what is represented and
> how. They will give rise to debate about the documentation and portrayal of

people and places. **Tenth** Deep maps will be unstable, fragile and temporary. They will be a conversation and not a statement.

Reproduced from: <http://cliffordmclucas.info/deep-mapping.html> [accessed 10 April 2020].

45 As quoted in Jones and O'Neill, *Living Maps*, 116.
46 Jones and O'Neill, *Living Maps*, 120.
47 Sylvia Loeffler, 'Place Values – *Glas Journal*: A Deep Mapping of Dún Laoghaire Harbour (2014–16)', in *Landscape Values: Place and Praxis*, ed. by Tim Collins, and Others (Galway: Centre for Landscape Studies NUI Galway, 2016), 30.
48 Quoted from the call for papers for the *Beyond Interdisciplinarity: Situating Practice*, Panel of the History and Philosophy of Geography Research Group at the 2016 Royal Geographical Society conference.
49 Elizabeth Young-Bruehl, *Hannah Arendt: For Love of the World* (New Haven, CT: Yale University Press, 1982), xxxi–xxxii.
50 Denis Wood, 'Mapping Deeply', in *Deep Mapping*, ed. by Les Roberts (Basel: MDPI AG - Multidisciplinary Digital Publishing Institute, 2016) pp. 15–29.
51 Denis Wood, *Everything Sings: Maps for a Narrative Atlas* (Los Angeles, CA: Siglio, 2011).
52 Wood, 'Mapping Deeply', 15.
53 Ibid., 17.
54 Ibid., 18.
55 Ibid., 16.
56 Ibid., 28.
57 See <http://www.lucigorellbarnes.co.uk/the-atlas-of-human-kindness/> [accessed 10 April 2020].
58 Tim Cresswell, *Place: A Short Introduction* (Oxford: Blackwell Publishing, 2004), 102–103.
59 Simon Read, 'The Power of the Ooze', in *The Power of the Sea: Making Waves in British Art, 1790–2014*, ed. by Janette Kerr and Christina Payne (Bristol: Sansom & Company, 2014), 46.
60 It is worth pointing out here that this word, 'resilience', has one of the most diverse applications of meanings imaginable across various disciplines. Used in psychology, for example, it has widely divergent associations than in anthropology, and these differ when used by a linguistics scholar. In negotiating scholarship and research across disciplines as they are currently demarcated in the academic world, understanding key terms is often a revelation and a necessary beginning.
61 Quoted in Brad Evans and Julian Reid, *Resilient Life: The Art of Living Dangerously* (Hoboken, NJ: Wiley, 2014), 71.
62 Jane Bailey and Iain Biggs, '"Either Side of Delphy Bridge": A Deep Mapping Project Evoking and Engaging the Lives of Older Adults in Rural North Cornwall', *Journal of Rural Studies*, 28, no. 4 (October, 2012), 318–328, doi:10.1016/j.jrurstud.2012.01.001.
63 Roberts, 'Preface: Deep Mapping and Spatial Anthropology'.
64 Erin Kavanagh, 'Re-thinking the Conversation: A Geomythological Deep Map', in *Re-Mapping Archaeology: Critical Perspectives, Alternative Mappings*, ed. by Mark Gillings, Piraye Hacıgüzeller and Gary Lock (London: Routledge, 2019), 201–230.

4   A call to action: reclaiming habitats through creative eco-social collaborations

Introduction

There is nothing more urgent for us humans to do than value all species and attend to the need for preserving; revitalising; and, in some cases, creating healthy habitats for our fellow creatures.

So says Christine Baeumler as she advocates for moving away from anthropocentric perspectives. "Loss impels action – how can I as an artist work with others to at least stem the tide or reverse the situation if possible?", she asks (Figure 4.1).[1]

Indirectly, grieving is a process that has motivated her work. Rather than lingering to do something about the grief she feels at ecological devastation, she is a woman of action. For example, she had the children in her recent Buzz Lab view a film called *The Vanishing of the Bees*.[2] "I thought at first I had made a terrible mistake", she said with a wry smile, "they were horrified, and looked stricken after they saw this. But there they were the next day, more determined than ever to do something about establishing a habitat for pollinators in the city".[3]

This is one small typical story that Christine shares about her eco-social activities. It is her intense power of conviction and overwhelming sense of empathy that moves people around her to action. In a recent catalogue of Christine's work, curator Mary Jane Jacobs remarked:

> . . . If we are more conscious, more mindful, what might the world look like? Baeumler thinks about this, but even more so she is *mindful* of this. She knows also that consciousness exists beyond the human mind and irrespective of us. There is an ethic in keeping this *in mind* – for the benefit of others.[4]

The eco-social

Christine Baeumler has moved away from conventional notions of the artist as singular maker, and instead moves towards enabling interdisciplinary teams and wider communities, including non-human beings (such as insect

Figure 4.1 Rooftop Tamarack Bog, Minneapolis College of Art and Design, 2012-
ongoing. Photo: Rik Sferra, 2012.

communities), to effect change. This approach to collective work is based on ecosophical understanding and engagement, using a geopoetic, metaphorical thinking attuned to the multiple meanings and contexts of the lived experiences of both human and non-human beings. Working as a public environmental artist, educator, curator and community activist, her actions work as a kind of catalyst to increase and facilitate awareness of environmental issues. In doing this, she aims to foster learning among and between many different types of beings: human specialists, youth, special interest and community groups and a wide range of non-human beings to facilitate appropriate responses to humanly damaged environments. Her work typifies the multidimensional, conversational approach explored elsewhere in this book, and also extends that mindfulness and dialogic listening to the non-human world.

Christine's approach incorporates both perspectives provided by art and the natural sciences, and social actions intended to engage with the complexity of ecosystems. She works to inspire creative solutions in order to understand and address species under threat by making room for natural habitats in the urban environment. If we can agree that we live in a world of contested spaces, and that humans all too often see mostly their own needs as primary, then the challenging dilemmas we face by imagining alternatives must reach beyond conventional approaches. Therefore, her practice is site- and community-based and requires her to consider the historical, cultural, ecological, metaphorical and aesthetic dimensions specific to each place while attending to pressing environmental and political issues that are in constant flux.

Collective ecological restoration of urban green spaces is necessarily collaborative, and trans-disciplinary in its need for the broadest possible understanding and implementation. By paying particular attention to increasing

Figure 4.2 Bruce Vento Nature Sanctuary, Saint Paul, Minnesota, 2012. Photo: Christine Baeumler.

biodiversity, habitats can be improved for plants and animals, as well as the water quality and aesthetic dimension of the sites concerned. Over the years, her art practice has gradually transformed from a conventional emphasis on an individual vision to the long term, engaged, collaborative and collective endeavours of a 'slow' place-based practice.

One example of this slow and sustained commitment may be traced to the *Bruce Vento Nature Sanctuary Project* (Figure 4.2). Back in 1994, as a resident and local artist, Christine began working on a series of community-led ecological restoration initiatives on the East Side of Saint Paul, Minnesota. The *Swede Hollow Historical Forest*, *Swede Hollow Henge*, the *East Side Gateway Rain Garden* and the *Maria Bates Rain Garden* were four distinct projects that were accomplished in collaboration with local residents, ecologists, hydrologists, engineers, University of Minnesota art students, the Como Park Conservatory Youth programme and the East Side Youth Conservation Corps of the Community Design Center. As a member of the Friends of Swede Hollow Park and a founding member of the Lower Phalen Creek Steering Committee since 1997, Christine worked with other East Side community activists on a grassroots effort to remediate a 27 acre, heavily polluted rail yard running along the Mississippi into the entirely transformed Bruce Vento Nature Sanctuary, a new Saint Paul City Park. As further testimony to this kind of slow and sustained commitment to realising long-term projects, she is also a core partner member of *Healing Place*, a local artist-led community project founded by Dakota artist and activist Mona Smith, which explores healing connections among people and place.

As the Artist-in-Residence in the Capitol Region and Ramsey Washington Metro Watershed Districts, she worked with these units of government on large-scale water infrastructure projects with the intention of making water quality issues and infrastructure more visible, educational and aesthetically compelling.[5]

Inspired by the possibilities of water infrastructure as an artistic form, Christine then led a team that included engineer Kurt Leuthold and ecologist Fred Rozumalski.

Together they created the *Tamarack Rooftop Restoration*, a 'bog ecosystem' above the entryway to the Minneapolis College of Art and Design (see Figure 4.1). This installation calls attention to these fragile and unique ecosystems, becoming in miniature a replication of the type of boglands so typical of much of Minnesota, as well as presenting an artistic re-imaging of a green roof. Seeing value in what were formerly termed 'waste' lands, she instinctively knew that they were not 'empty' (Figure 4.3).

Reclaiming and revaluing are key words that typify Chris' practice. She is mindful of species other than human, and has been consistently moving away from anthropocentric perspectives over the last 25 years in her teaching and her practice. Every project, every research action, has shown her the importance of framing the questions as a matter of healthy and respectful

Figure 4.3 Itasca Community College bog, from *Bogs, A Love Story, Documentary*, 2012. Photo (video still): Christine Baeumler.

Figure 4.4 Pollinator Garden, Plains Art Museum, 2017. Photo: Christine Baeumler.

co-existence, which she says "benefits us all". Certainly, it is healthier from the perspective of all the other species that have been so decimated in recent decades by negligent damage to their habitats.

Christine proposed, and is the lead artist on, the project called *Pollinators at the Plains*, which is a sustainable redesign in Fargo, North Dakota of the Plains Art Museum's outdoor campus (Figures 4.4 and 4.5). This included a youth internship program as an essential part of the project and involved activities to educate youth about pollinators, with artist and horticulturist Seitu Jones leading the installation of a fruit orchard (Figure 4.6). The interns associated with this project did a series of art projects and performed at the opening of the exhibition, *Living As Form, the Nomadic Edition.*

Similarly retaining an element of nomadic journeys in her own life, Christine has an artistic practice that exceeds the bounds of the Midwestern USA. In addition to the fact that she is committed to local and regional site-based projects, she has also done temporary projects in Chengdu, China addressing water quality (*Dreams of a Pure River*, 1995) and work in Germany (*Bureau of Atmospheric Anecdata*, 2008), both of which were collaborations with California-based artist Beth Grossman. She is a member of the international research networks PLaCE International and Mapping Spectral Traces, whose work, often intensely local, addresses issues of environmental

Figure 4.5 Swallowtail Butterfly, Pollinator Garden, Fargo, North Dakota, 2017. Photo: Christine Baeumler.

and social concern that overlap with her own. And her most recent research, with Mary Modeen, has been in Ilhabela, Brazil.

Christine has also organised symposia and co-curated exhibitions as a means to create platforms for artists and scholars and so foster a community of practice around engaged collaborative scholarship. Her co-curated exhibitions at the Katherine E. Nash Gallery at the University of Minnesota included: *Critical Translations*, *Mapping Spectral Traces* and *thinking, making, living*, a four-month long exhibition that introduced students to local and nationally recognised socially engaged artists.

Baeumler's practice begins with a simple premise: revaluing as a first step in the creative initiative. She believes that one can make anew from the remains of the old, the imbalanced, the cast aside and the forgotten. It suits her conscious decision-making steps to believe that socio-ecological and community-political steps can coincide by revitalising sites in simple steps. In setting this out as an action plan, she begins stages of thinking about a long-term project, initially motivated by a sense of loss or urgency. This is not just activism (or informing through metaphorical thinking as an artist) but in a wider sense combines art and activism. Through this, she

Figure 4.6 Buzz Lab youth interns with artist Seitu Jones planting orchard at Plains Art Museum (2014). Photo: Cody Jacobson.

believes, one engages more simultaneous 'poetic layers' of action, because it draws upon other people's imaginations, allowing them space to participate and even initiate their own independent actions. It opens up agency for others, including non-human others (Figure 4.5) leading to larger collaborative actions. She understands that this type of work "is about transforming other people's perception, shifting it, in ways that may just create a culture shift", she says. "By interviewing 500 people about their stories, the collective power of many stories convinces them to believe that they have stories that matter – it becomes a power to shift an entire cultural attitude and perception". The nature of this artistic concept was cited by Mary Jane Jacob, as quoted above, but also echoes John Dewey, in his essay about experience and knowingness. Dewey wrote:

> The statement that individuals live in a world means, in the concrete, that they live in a series of situations. And when it is said that they live *in* these situations, the meaning of the word "in" is different from its meaning when it is said that pennies are "in" a pocket or paint is "in" a can. It means, once more, that interaction is going on between an individual and objects and other persons. The conceptions of *situation* and *interaction* are inseparable from each other. An experience is always what it is because of a *transaction* [italics added] taking place between an individual and what, at the time, constitutes his environment.[6]

This notion of the transactional, then, as articulated by Dewey and shared by Baeumler, is central to her commitment to the eco-social for reasons of the awareness of movement between the situation and the interaction.

Her practice, her words

In the following section, Christine Baeumler (**CB**) responded to questions put to her by us, Iain Biggs (**IB**) and Mary Modeen (**MM**). For the sake of separation, the interviewers' questions are in italics, and Baeumler's answers are in regular text.

IB: *You've always avoided narrow categorisations of your practice. You're interested in raising awareness of ecological issues through studio practice – whether painting, printmaking, film, installation, or whatever – and at the same time engaging communities in environmental projects focussed on ecological restoration. This work with your local community seems central to so much of what you do. Why is this multiplicity of approaches important to you?*

CB: My practice has shifted over the years from a studio-based work (paintings, drawings, photographs, video and multi-media installations about endangered species and threatened places) to community-based site-specific projects. Currently, I am exploring a more systems-based approach, with an emphasis on storm water treatment through a green infrastructure and habitat restoration, particularly for pollinators. Considering that ecological systems are interrelated, I ask questions about how an artistic practice can make complex systems visible, comprehensible, aesthetic, interactive and functional – while avoiding didactic solutions.

It might be useful to say something about my background. I first recognised the power to bring a diverse group of people together around a mutually held mission when I volunteered on an urban farming and historical project in East Palo Alto, California in 1990. The organisation, the East Palo Alto Historical and Agricultural Society (EPA–HAS), was founded and led by artist and activist, Trevor Burrowes. As a board member of EPA–HAS, I learned the role artists could take by being embedded within a non-profit organisation and within a community context. The activists, urban farmers and community members in East Palo Alto rallied around a common passion for local history, healthy food, youth involvement and organic farming (Figure 4.7). The efforts resulted in the development of a farmer's market in a community that was once a food desert. While urban farming has caught on in recent years, this organisation's vision was well ahead of its time. I witnessed first-hand how environmental and social goals could bring people together from different cultural and economic backgrounds. It was a powerful lesson of transformation accomplished by a small group of people with a vision – and this experience reshaped the trajectory of my artistic path.

Figure 4.7 Leroy Musgraves, a founding member of East Palo Alto Historical and Agricultural Society at his urban farm in East Palo Alto, California, 1993. Photo: Christine Baeumler.

While in California, I met three key women artists, Betsy Damon, Rhoda London and Beth Grossman, through the 'No Limits for Women Artists' organisation, a group dedicated to the promotion of women's creative activities. Betsy Damon was the founder of 'No Limits' and was living in Minnesota when I moved to Saint Paul in 1994. She also founded the 'Keepers of the Waters' organisation and in 1995 invited me (in collaboration with artist Beth Grossman) to do a temporary socially engaged project in Chengdu, China, about the water quality of the Fu and Nan Rivers. Upon my arrival to the Twin Cities, Betsy also introduced me to the recently formed 'Friends of Swede Hollow', a grassroots neighbourhood organisation that formed to revitalise Swede Hollow Park in St. Paul.

I discovered quite early on that working with specific sites involved the geological, hydrological, ecological past and present conditions as well as the cultural dimensions of the site. Your notion, Iain, of the deep-mapping of a site immediately resonated with me in terms of the consideration of the multiple dimensions involved when one works in a place-based situation. The studio art and site-specific community-based work felt like two parallel tracks for a while, and though I knew of other artists who were working ecologically, initially it was difficult for people to recognise the ecological work within the frame of an artistic practice. There was little recognition or support (in the academic, arts, or funding communities) that this collaborative and community-based work could fall within the accepted definition of 'art'. Estella Conwill Majozo's call to action inspired me during this time: "To search for the truth and make it matter: this is the real challenge for the artist. Not to simply transform ideas and revelations into matter, but to make those revelations actually matter".[7]

IB: *The historical context is obviously important here. Do you think official attitudes to approaches like your own are changing and, if so, why?*

CB: Looking back, the Bay Area in Northern California during the 1990s was one of the places in which artists and educators initiated and began to make legible 'New Genre Public Art'. Critic Suzi Gablik describes this practice as involving:

a distinct shift in the locus of creativity from the autonomous, self-contained individual to a new kind of dialogical structure that frequently is not the product of a single individual but is the result of a collaborative and interdependent process.[8]

Most of my own formal art education during the 1980s emphasised both formal and expressionistic approaches – the legacy of the Bauhaus and Abstract Expressionist attitudes that prevailed in university art departments in the 1980s and 1990s. Given the emphasis on the expression of an individual artistic vision and a market-driven model of success, a more collaborative approach was in direct contradiction to any definition of art as I was taught it.

As you are well aware, 'socially engaged art' or 'social practice' has recently emerged as a recognised field in the arts and academia. A growing number of universities are offering programs and degrees in social practice, while curators are mounting exhibitions dedicated to these areas as more closely articulated foci. A more substantial body of literature and critique has developed over the past two decades. The variety of approaches has made it difficult to create a single umbrella term that satisfies anyone.

Artists, critics, theorists and curators are all grappling with the spectrum of this practice and a critical discourse has emerged around the complex ethics of working on projects within a community context. A new generation of younger artists is emerging who are challenging the strategies of the past two decades or so, and are currently redefining what it means to be socially engaged through the creation of participatory platform and networks. So perhaps it is for the benefit of this practice that it remains perpetually in flux and that we continue to challenge each other as the field continues to evolve.

Just as the boundaries between the various media within art are starting to break down, I also see the disciplinary boundaries between art and many other disciplines starting to erode. It was interesting to attend a geography conference a few years ago and witness so many artists present their work in the context of a place-based discipline. In my public projects, I have worked on teams with ecologists, engineers and hydrologists. This work truly is collaborative and could not exist without the expertise of these team members (Figure 4.8).

Figure 4.8 Pollinator Skyrise, collaboration with artists Amanda Lovelee and Julie Benda and Blue Rhino Studio fabricators, Como Park, Saint Paul, Minnesota, 2017. Photo: Christine Baeumler.

One thing that has fascinated me in recent years is recognising the particular contribution – the ecological niche in large community and government-sponsored projects – and the facilitating role that I play as the artist. I am not a designer but rather someone who advocates opportunities for experiences that are prompted by the features of the site, without being overly programmed.

The other part of this process that has fascinated me is that during project discussion, the experts move outside of their own disciplinary areas to make suggestions about areas that are not in their own wheelhouse. The projects are most exciting when they emerge in an organic way and a collective approach to idea development takes over.

My feelings are somewhat mixed about 'institutionalising' this type of practice – I think part of its vibrancy is precisely in its community-based roots outside of academia. I have some reservations about teaching 'social practice' as I think students and artists need to come to this practice through a passionate concern about an issue, to approach art as a catalyst for social transformation. So, it seems somewhat problematic to adopt 'socially engaged' art as a methodological practice without a strong reason to pursue an instrumental approach to one's art.

Figure 4.9 Buzz Lab fieldwork at Edward M. Brigham III Alkali Lake Sanctuary in
Spiritwood, North Dakota, 2017. Photo: Chris Baeumler.

In my Art and Ecology course, I am taking my students out in the 'field'
rather than spending most of our time in the classroom. We will be meeting
artists and activists on site to have a direct dialogue and observe the projects
in situ (Figure 4.9). The site specific, socially engaged projects I worked on
starting in the mid-1990s emerged from working collaboratively with people
from a variety of fields to reclaim urban ecosystems. One of my primary aims
is to reconnect youth and residents to nature in their own neighbourhoods
through the revitalisation of degraded green spaces and the replacement
of manicured turf. Collaboration, facilitation and team building became
essential ways of working.

The feminist and emerging environmental artists of the 70s, 80s and 90s
provided an eye-opening counter narrative – but I only became more fully
conscious of this work after undergraduate and graduate school.

IB: *Did your approach evolve organically over time through the practice itself
or has it been a conscious engagement with particular theoretical positions?*
CB: I believe that my practice evolved organically along with the arts community in Minnesota through a kind of ecology of practice that includes
not only artists, but also community activists and non-profit organisations. Overall, I've been more inspired and influenced by artists as

models rather than theoretical frameworks. I feel more closely akin to the idea of praxis – ecological and ethical concerns put into practice. I felt inspired early on by John Dewey's notion of art as experience – and by the many artists that challenged the notion that art should be prescribed by strict boundaries and defined in narrow terms.

When I started this work 20 years or so ago, there was no clear roadmap for ecologically oriented artists. For me, it has been more helpful to see how artists have applied these ideas and manifested them in the world – with all the messy complications of human relationships, city ordinances, financial constraints – rather than a more abstract sphere of theoretical positions. Artists such as Josef Beuys, Mierle Laderman Ukeles, Mel Chin, Alan Sonfist, Newton and Helen Meyer Harrison, Lorna Jordan, Jackie Brookner and Buster Simpson gave me a kind of permission and support through their own work that what I do can actually fall within the realm of an artistic practice. I benefitted from the direct mentorship of environmental artist Betsy Damon. I also collaborated with Bay Area artist Rhoda London for a number of years on installations about ecological issues, particularly about the extinction of bird species. And I engaged in a number of environmental and socially engaged collaborations with Bay Area artist Beth Grossman and with Minneapolis artist Kevin Johnson. My own practice has evolved with what is now called 'Social Practice'. While I was not in the initial wave of feminist artists, I feel powerfully influenced by them – women that in many ways are not getting the attention they deserve.

Early on, I recognised the dizzying complexity of place-based engagement, and how necessary it was to work in this way if one wanted to work with ecological systems and cultural dimensions. It is not an overstatement to say that one simply *had* to work with a trans-disciplinary mind-set in order to make progress. I also realised quickly that I had to collaborate with people with a depth of knowledge from a variety of disciplines – that I couldn't possibly be an expert in all these fields – and also with people with experience and a different cultural perspective that sits outside disciplinary thinking. I have been re-oriented over the years by people like artist Mona Smith and science educator Jim Rock, both Dakota, who assisted me to understand indigenous perspectives and at moments, at least to shift the sense of my relationship to the river, the land – and even the stars.

Working as the Artist in Residence for two Twin Cities watershed districts has shifted my approach from thinking about a particular site to considering larger inter-related systems: water infrastructure as a larger system that involves public and private sites of ownership as well as the processes of precipitation, storm water runoff and evaporation. It struck me when I heard Chad Staddon use the term 'Hydro-social' (originally coined by Manchester geographer Erik Swyngedouw) at the Catchment Conference in Bristol, UK in 2010. The watershed districts in Minnesota are small units of government working at the intersection of both hydrological and social solutions to

Figure 4.10 The 'Climate Chaser, *Backyard Phenology Project*,' a mobile phenology lab and sound studio, 2016. Photo: Christine Baeumler.

water quality. I saw the potential of art to be a bridge between the hydrological and social approaches to water quality issues in our community.

MM: *Can you tell us something about the current Backyard Phenology project?*

CB: The current Backyard Phenology Project has emerged as a revelation (Figure 4.10). The shift in register from the sole observer to a collection of a whole community of 'citizen-scientists' has helped me see a larger way to connect eco-social actions beyond the individual to move toward a meta-data collection, while at the same time maintaining the importance of each individual with their own enthusiasms, their own minute observations and growing tenderness for their back gardens in ways that hadn't included in this level of scrutiny or sharing before. The seasons, tiny examples of climate change made visible, daily observations – this citizen science project has brought together collected information where individual observations become a valuable part of 'big data'.

One of the Backyard Phenology Project's host sites is the Native American Medicine Gardens at the University of Minnesota. The concept of 'phenology' is related to the Lakota perspective of 'observing as you go', translated as *akita manni yo*, as shared by *Cante Suta* – Francis Bettelyoun, the coordinator of the Native American Medicine Gardens, who is Oglala Lakota.

This perspective would remind us always to 'walk in this world with eyes wide open'. As a landscape designer and native master gardener, Bettelyoun has significantly informed the Backyard Phenology project about the significance of integrating native cultural perspectives in all aspects of life related to the natural world as well as the importance of acknowledging that the land in Minnesota is that of the *Očhéthi Šakówiŋ*, the Seven Council Fires which include the Dakota, the Lakota and Nakota.

IB: *You're interested in careful scientific observation of the natural world, particularly as this relates to understanding the idea of non-human sentience and of our connection to other species. This would seem to link very clearly with moves to create a greater permeability between art-based and scientific approaches?*

CB: Over the years, I have approached both my studio and public art practice through the combined perspective of art and the natural sciences. In the studio practice, I have often placed myself in the role of the careful observer and have shadowed the journeys of earlier naturalist explorers who travelled to investigate unfamiliar species and habitats. I've been fortunate to travel to the Northern Australian and Amazon Rainforests, the Great Barrier Reef and the Galapagos Islands in Ecuador (Figure 4.13). I often accompanied scientific research teams and tours guided by naturalists so that I could, at least temporarily, look through the lens of science. The paintings, installation and the video work departed from scientific inquiry in terms of the subjectivity of the results – it exists not as a description of place but rather a record of my experience through the process of perception and memory (Figure 4.11).

My concern lies not only with the diminishment of these species and ecosystems but also with the extinction of the human experience and knowledge of these environments and the species that inhabit them. By portraying worlds remote from our daily experiences, yet impacted ecologically by our actions, that particular body of work was intended to offer the viewer a glimpse into these compelling, fragile and often invisible worlds.

In both my studio and public practice, I have focussed on the notion of making invisible systems visible. My intention in the urban ecological reclamation work, whether it is a rooftop bog, or a rain garden for pollinators – is that city dwellers will make room for other species and, optimistically, become better stewards of the ecosystems they inhabit on a daily basis. Much research has been done on the benefits of green space to human well-being. So, there are multiple and reciprocal benefits to creating more biologically rich urban green spaces.

IB: *How does this sit with your perception of the current state of art and education?*

CB: I'm excited to have been around long enough to see cracks form and outmoded disciplinary silos start to blur and break down. Over the

Figure 4.11 Darwin's Table: installation at the Phipps Center for the Arts, Hudson, Minnesota, 2010. Photo: Christine Baeumler.

course of my career, I have witnessed a shift in how artistic practices are 'siloed', both within academic and art institutions as well as in the funding world. Our art department – like many in the US and elsewhere – are carved into discreet media disciplines. Art faculty themselves have been working in an inter-disciplinary and intra-disciplinary ways for a long time – but the systems in which art education has been taught is much slower to shift due to the way people are hired, the management of facilities, the way graduate students are admitted and so forth. Many of us in academic settings took the 'fly under the radar' strategy in programs that were still conventional in their structure. To our own surprise, perhaps, engagement with communities and a growing interest in environmental concerns has begun to value the work of those who have insisted on working locally or outside of the art market. Similarly, the way art funding has been set up has also been structured on this idea of discreet and often traditional media definitions. It's worth noting that artists have been on the leading edge of reimagining and defining their own practices.

While many artists in the Twin Cities have been working in an interdisciplinary, community-based and site-specific way, I have to also give credit to the visionary curators and leaders of local non-profit arts organisations in my region.[9] All of these people have been extremely instrumental in legitimising these contemporary practices. Each has been deeply committed to artists who didn't fit into the prevailing system of the art market and has created opportunities with an openness that allowed artists to lead interdisciplinary groups.

IB: *Could you tell us more about your project 'Reconstituting the Landscape: A Tamarack Rooftop Restoration'?*

CB: The first challenge was that people want to remove water from a roof, and our proposal called for an open water feature as part of the design. Kurt Leuthold designed a closed system in which the rainwater collects on the roof and eventually falls into a cistern below. A solar powered pump recirculates the water back onto the roof to keep the thirty-two plant species and tamarack trees watered. Transpiration from the plants also releases water into the atmosphere.

The plants themselves were 'salvaged' as they were removed from an area in Northern Minnesota to make a pathway for a utility line – they were displaced from their original environment and 'rehomed' on the roof. Ecologist Fred Rozumalski worked closely with Boreal Natives who rescued the plants, and then grew the plants in mats in a greenhouse before their installation.

Luckily, I had a determined advocate in the gallery director, Kerry Morgan, who believed in the project, as well the support of a local watershed organisation and local funder, in addition to the McKnight Foundation. It was a good example of how many allies it took to establish a 16 by 24-foot ecosystem! The bog has been up now since 2011 and going into its ninth season. In reality the project extended beyond our three-person team to include the gallery director, the facilities manager, the roofers and the landscape staff at Boreal Natives who collected and helped to install the plants. The security guards have become informal guides when visitors arrive – I have overheard them talking to visitors about the bog. This is all very satisfying to me since there was quite a bit scepticism when I first proposed the idea.

IB: *The project seems to raise a number of interesting questions. For example, you seem to be artfully engaged in trying to 'reframe' how we think about ecologies; to provide other models and orientations in ways that relate quite closely to issues raised by George Lakoff around communicating environmental issues.*[10]

CB: I believe this reframing is a critical step in changing our attitudes and our actions towards what we consider the natural world. Timothy Morton challenges us to relinquish an historical notion of Nature as something separate from us.[11] Morton's idea of 'the mesh' requires us to abandon a notion of nature as something apart from us and come to recognise the importance of how things are inter-related. He writes:

Ecology shows us that all beings are connected. The ecological thought is the thinking of interconnectedness . . . It's a practice and a process of becoming fully aware of how human beings are connected with other beings – animal, vegetable or mineral. Ultimately, this includes thinking about democracy.

What would a truly democratic encounter between truly equal beings look like, what would it be – can we even imagine it?[12]

My shift to an approach aligned more to whole systems, and moves beyond site specific considerations to study the ways specific places, species and systems are inter-related. It's urgent we are able to think about larger, complex systems and how we – and in turn these systems, such as climate – have impact not only on ourselves but on the lives and very existence of other species as well.

MM: *This sounds a lot like your thoughts about the more recent Backyard Phenology Project, as described above.*

CB: I believe that it is important to engage many people as observers rather than to rely upon a sole observer, who records, issues statements and makes proclamations as in the past. A collective observation that empowers lots of people has the effect of becoming much more attuned to the population, at the same time as drawing in people who learn more about their immediate environment through close scrutiny; we can't deploy experts in every back yard, but each homeowner can now become an expert. She/he can truly understand the effects of Climate Change through direct observation in the smallest of details.

IB: *So, what kind of experience do you hope to offer to the viewer?*

CB: I hope that a visitor to MCAD might be ambushed by the bog roof installation – an unexpected landscape in an unlikely place that provokes curiosity. The overall outcome of this piece is an unusual semi-permanent roof garden that calls attention to the unique ecosystem it mimics and also models a re-imagining of the whole idea of a green-roof infrastructure. This might prompt students and the general public to think about how an eco-system such as a bog can be 'reconstituted' using rain water capture to maintain itself, and spark ideas about how it can function as a way to manage storm water. For me, it was also interesting to think about how it functions as discreet landscape bounded by a rectangle. It is only accessible visually – it creates a little miniature world unto itself – a 'bonsai bog', and can be viewed from below but also from the second floor and on the skyway through the tall vertical glass windows of the existing architecture. It struck me one day that the rooftop bog was really not far from my first impulses to create terrariums as a child or later to paint the landscape – and an attempt to have a fragment suggest a whole world unto itself, the metonymy transporting the viewer into larger realms through the observation of a fragment.

IB: *So how does this modelling of a possible reconstituted landscape relate to your ongoing interest in environmental issues more generally, particularly those around adaptation and extinction?*

CB: The rooftop bog has deepened my understanding and commitment to bringing attention to the value of bogs and swamps. Through my research, I met a number of passionate 'bog people' from a variety of disciplines—and was intrigued that these rather inhospitable places inspired such fascination and devotion (Figure 4.12).

Bogs and peatlands are alternately considered desolate wastelands or vast untapped natural resources. In the upper Midwest, as in many places around the world, peatlands have been drained for agriculture land, stripped of their existing lumber and mined for their peat as fuel and as a horticultural product.

In *Stirring the Mud: On Swamps, Bogs and Human Imagination* Barbara Hurd notes:

> To love a swamp, however, is to love what is muted and marginal, what exists in the shadows, what shoulders its way out of mud and scurries along the damp edges of what is most commonly praised. And sometimes its invisibility is a blessing.[13]

Largely unrecognised and misunderstood is the role bogs and peatlands play in the climate change equation. These landscapes provide significant

Figure 4.12 (left to right): Laura Hope, Maris and Chris Baeumler at Rannoch Moor, examining bog plants *in situ*. Photo: Mary Modeen, 2016.

yet rarely appreciated ecological benefits. Beyond providing habitat to unique plant and animal communities, peatlands cover approximately 3% of the earth's surface and yet are a carbon sink for 30% of its carbon dioxide. What happens in bogs also impacts water availability. Scientists are investigating now what happens if extreme events grow yet more extreme and become more frequent. Are there more frequent droughts likely to occur in the near future? Are there more frequent high-severity windstorms or rainstorms ahead? These events have both ecological and cultural implications for communities in terms of water resource availability.

In the summer of 2014, I had the opportunity to have an exhibition at the MacRostie Art Center. For that installation I produced *Bogs, A Love Story*, a short film documentary about six different bog experts: a photographer, Elizabeth Blair, who is an expert on orchids; John Latimer, a former mailman and phenologist; Stephen Sebestyen, a climate scientist; Bill Marshall a former regional land manager; Julie Miedtke, a forester/educator; and Harry Hutchins, an ornithologist. I discovered through the making of this short film that the Marcell Experimental Forest (MEF) in Grand Rapids, Minnesota, is home to the largest climate change experiment on peatlands in the world: it is the Spruce and Peatland Responses Under Climatic and Environmental Change Experiment (SPRUCE).

I aim to form now an interdisciplinary team to document the SPRUCE project as it provides a once-in-a-lifetime opportunity for artists, designers and community members to work alongside scientists on the front lines of climate change research, at the beginning of a ground-breaking ten-year study. Our proposed project further catalyses the local community's creative capacity and imagination to establish new narratives around a hydro-centric climate future, which will be shared with peatland communities worldwide, who will also be impacted by climate change (Figure 4.13).

IB: *You've spoken in the past of the privilege of being in close proximity with particular species – certain birds, whales and sea lions – and of how the sense of reciprocity evoked in that situation was transformative. Can you say something about this – both in terms of the transformative experience itself but also about whether you think engagements with animals and landscapes through art can evoke similarly transformative experiences?*

CB: I have snorkelled amongst dwarf minke whales, sea turtles, barracudas, lionfish, reef tip sharks and an incredible array of fish at the Great Barrier Reef and floated amongst sea lions, Galapagos penguins and hammerhead sharks in the Galapagos. In the Amazon, I joined scientists through an Earthwatch expedition to count populations of pink and gray river dolphins and macaws and, in the Amazon jungle, to count the red uakari monkeys overhead. Through these encounters, I became keenly aware that not only was I the observer, but that I had also become the subject of the encounter – that I was, temporarily, a part of that ecosystem – and the one being observed!

Figure 4.13 Chris Baeumler recording a marine iguana, Galapagos, 2007. Photo: Jonathan Wells.

I have long been inspired by Simone Weil's notion of 'de-centring', described in Elaine Scarry's book *On Beauty and Being Just*, in which one has to step out of the centre of one's own world to make room for the Other – while the space created by the removal of the self makes room for empathy to enter. She says:

> It is not that we cease to stand at the center of the world, for we never stood there. It is that we cease to stand even at the center of our own world. We willingly cede our ground to the thing that stands before us.[14]

When we apprehend something beautiful, we no longer stand at the centre of ourselves. I experienced first-hand a kind of radical de-centring – submerged in the ocean and interacting with the species around me. I experienced myself as just a part of a greater whole, along with a profound sense of reciprocity as the one seeing and being seen, listening and being heard.

Suzy Gablik states that:

> empathic listening makes room for the Other and decentralises the ego-self . . . Art that is rooted in a 'listening self', that cultivates the inter-twining of self and Other, suggest a flow-through experience which is not delimited by the self but extends into the community through modes of reciprocal empathy.[15]

These profound encounters with other species deepened my understand-ing of the importance of opportunities for people to continue to connect to actual plant and animal species, rather than to representations in media. As I continue to consider urban and suburban greenspace as ecological corri-dors and habitat, particularly for pollinators, I imagine that anyone can have a moment of an encounter, with a monarch butterfly, a pileated woodpecker, or a chipmunk. I don't think one has to travel long distances to see 'exotic' species to have that experience of a connection, but simply to be attentive to one's own environment through all the senses. So, I think of the current work as creating increased opportunities for those inter-species moments of contact as well as beneficial habitats. I am currently working on several new projects that focus on the development of urban pollinator habitats.

MM: *What are your aims for the Pollinator Garden at the Plains Art Mu-seum in Fargo, North Dakota, in relation to observation and inter-species contact?*

CB: *The Defiant Gardens for Fargo-Moorhead* is a public art project for the twin towns of Fargo, North Dakota and Moorhead, Minnesota, and is intended to create a diffused network of gardens integrated into unex-pected sites in the community. The project was developed by the Plains Art Museum, located in Fargo. Art Museum director and CEO Colleen Sheehy, conceptualised the original idea for the project and served as project director.

The theme of my 'defiant garden' proposal was to create a site that would provide habitat for urban pollinators while educating residents, and par-ticularly youth, about the significance of pollinators and the current threats to their survival. Considering that one out of every three bites of food we consume is produced with the assistance of pollinators, one would imagine that species so central to our own diets would be more prominent in our consciousness. Yet most people are unaware of the significant role pollina-tors play in sustaining the structure of the landscape and our food system; nor are most people aware many pollinators indigenous to North America, for example, have been in a precipitous state of decline in recent years due to a variety of environmental factors.[16]

In this community project, we also installed a cistern to collect water from the apartment building adjacent to a parking lot that automatically waters the orchard trees. The large cistern is painted with iridescent colours for the pollinators, based on drawings created by the Buzz Lab interns. The landscaped plants and trees were planted during the summer and fall of 2014 with the Buzz Lab interns. When I initially proposed the Pollinator pro-ject, I was clear that I wanted to include a youth internship and educational component as an essential part of the project itself. I also wanted it to reach a diverse youth population, as there are currently many families who have newly immigrated to the communities of Fargo/Moorhead from Africa and the Middle East. It was important that this work was to be a paid internship. For almost all the interns (all of whom were under the age of 18) this was their first paid job.

With the Plain Art Museum staff's assistance, we scheduled an entomologist, commercial and residential bee-keepers, a landscape architect, gardeners, poets and a chef to be guests for Buzz Lab. In the summer of 2014, the paid internship involved an intensive 40-hour week of activities to educate this youth team on types of pollinators and further sessions dedicated to edible and native plants as well as the fruit orchard, all under the direction of Seitu Jones. The interns also did a series of art projects including a Public Service announcement called 'Don't Fear the Bugs'.

Over the course of the Buzz Lab in 2014, the youth who were involved became informed and genuinely concerned about pollinators. It was inspiring to see young people shift their perceptions from a 'fear of the bugs' to a position of advocacy and stewardship (Figure 4.14). Perhaps one of the most gratifying outcomes of the Buzz Lab was the relationships that developed between the youth in the cohort, who were from diverse cultural and class backgrounds. In the summer of 2015 there were 21 Buzz Lab interns; eight of these were youth returning from the previous year. This group of social media-savvy youth started several media campaigns to raise awareness about pollinators and the benefits of local foods. This group has impressed me with their enthusiasm for insects – and weeding! Buzz Lab 2.0 has provided much-needed maintenance of the gardens. The team also performed a

Figure 4.14 Queen Bee and Mobile Hive performance, Buzz Lab interns, Plains Art Museum, Fargo, North Dakota, 2017. Photo: Christine Baeumler.

'Council of the Pollinators', in which each intern created a mask of a pollinator and spoke at a public assembly from the perspective of that pollinator.

MM: *How would you summarise your practice generally?*
CB: I would like to share the overarching tenets to which I adhere. These are:
  • To acknowledge tribal homelands[17]
  • To respect Indigenous knowledge and diverse ways of knowing
  • Reclamation of urban sites: an intention to remediate abandoned, neglected or marginal places
  • To re-evaluate ecological functions: thinking and continually re-assessing the less-than-obvious functions of component elements in a given ecosystem through close observation
  • To consider the aesthetic dimension in place-based research, but not to the exclusion of other concerns such as sustainable materials and approaches
  • To engage in collaborative work, with
    – Community and youth engagements
    – Interdisciplinary projects with engineers, hydrologists, ecologists, people with cultural knowledge
  • A commitment to long-term, embedded presences
  • To create urban habitats for non-human species

Over the years a set of guiding principles have emerged and consistently informed my practice:

Making the invisible visible
Paying attention
Witnessing
Reciprocity
Recognition
Listening
Connecting
De-centering
Collaborating
Remaining unsettled
Transforming perception
Confounding expectations
Keeping the invisible
Invisible

Notes

1 Christine Baeumler, as quoted in an interview with the author, Mary Modeen, January, 2018. Hereafter, this will be referenced simply as 'Interview'.

2 *Vanishing of the Bees* is a documentary film directed by George Langworthy and Maryam Henein, and released in the United Kingdom by Hive Mentality Films & Hipfuel Films in October 2009. The 'Buzz Lab' is explained more fully below.

3 Interview, 2018.

4 Mary Jane Jacob, Executive Director of Exhibitions and Exhibition Studies at School of the Art Institute of Chicago, Director of the Institute for Curatorial Research and Practice at SAIC, *McKnight Visual Artists Fellowship Exhibition Catalogue* (Minneapolis, MN: MCAD, 2011).

5 Under the auspices of Public Art Saint Paul and supported by the Bush Foundation (2010–17).

6 John Dewey, *Experience and Education* (New York: Collier, 1938), 43.

7 Estella Conwill Majozo, 'To Search for the Truth and Make it Matter', in *Mapping the Terrain: New Genre Public Art*, ed. by Suzanne Lacy (Seattle, WA: Bay Press, 1995), 88.

8 Suzi Gablik, 'Connective Aesthetics: Art after Individualism', in *Mapping the Terrain: New Genre Public Art*, ed. by Suzanne Lacy (Seattle, WA: Bay Press, 1995), 76.

9 Colleen Sheehy, director of the Plains Art Museum; Christine Podas-Larson director, Public Art Saint Paul; Jack Becker director of FORECAST Public Art; Kerry Morgan, director of the Minneapolis College of Design gallery; Steve Dietz, director Northern Spark and Sarah Schultz, who until recently was the head of Public Engagement at the Walker Art Center.

10 George Lakoff, 'Why It Matters How We Frame the Environment', *Environmental Communication*, 4, no. 1 (2010), 70–81.

11 Timothy Morton, *The Ecological Thought* (Cambridge, MA: Harvard University Press, 2010).

12 Morton, *Ecological Thought*, 7.

13 Barbara Hurd, *Stirring the Mud: On Swamps, Bogs and Human Imagination* (Athens: University of Georgia Press, 2008), 8.

14 Elaine Scarry, *On Beauty and Being Just* (Princeton, NJ: Princeton University Press, 1999), 77.

15 Gablik, 'Connective Aesthetics Art after Individualism', 82.

16 Collaborators on this project have included Barr Engineering on the overall Master Plan, and Fred Rozumalski, who served as the ecologist. Baeumler worked closely with the Plains Art Museum staff and artist MeLissa Kossick (who became a co-teacher and leader over the course of the project) and artist Seitu Jones, who designed the orchard with the Buzz Lab. The implementation of the project is the culmination of five years of planning. In the summer of 2014, we worked with a landscape firm to remove pavement from the parking lot adjacent to the Plain Museum's Center for Creativity to create a series of planting beds for native plants and an area for an orchard designed to mitigate storm water runoff from downspouts which previously discharged rain water directly onto the pavement. The intention was to provide pollinator habitat as well as to grow vegetables, herbs and fruit for the staff and participants in the project.

17 In Minnesota where Christine Baeumler is based, these include the Seven Council Fires (*Očhéthi Šakówiŋ*) comprised of Dakota, Nakota and Lakota; Ojibway and Hochunk tribal nations also reside in Minnesota.

5 Perception and cultural memories of place

Dotted around the northern half of Scotland, in a *strath* (or valley) flanked by wilder, higher ground, there are small hillocks with clusters or rings of trees. These are usually Scots pine, standing tall and sharply silhouetted against constantly changing skies. The locals sometimes refer to these as 'Goodman's Lands'.[1] What they mean by this is slightly oblique and needs an explanation. These sites almost always contain prehistoric traces of human presence; they are usually the sites of burial cairns, or mounds, heaped with stones. A few have standing stone circles, and some have *howes* or 'keeps': ancient spaces lined with slabs of stone and covered over with earth. What the farmers know about these sites on their lands is capable of producing powerful emotions. They sense the traces and feel deeply the past presence of earlier men and speak (when they can be urged to speak about such things) of haunting or, as modern men, laugh a bit uneasily at superstitious fears (Figure 5.1).

Since at least the 18th century these places have been associated with the devil, but rather than inadvertently beckon The Dark One by saying his name, the locals say 'Goodman' instead.[2] Even in days of agricultural rationalisation, these circles of land will not be breached, remaining intact and undisturbed, leaving the Goodman his Land.

This chapter begins with this example of spectral traces because a traveller across Scotland's landscapes sees them every day, often without recognising them as such. They are stark images – beautiful and strange. They are also enigmatic in the way that land itself is enigmatic: to 'read' a place is to apply a process of interpretation, and to 'see' the landscape in itself as a story. This happens nearly simultaneously with the act of perception. In fact, depending upon whether one speaks to an artist, a philosopher, or a cognitive psychologist, each would describe this process of 'taking in' the world, and 'making sense' of it, deploying different language but arriving at similar conclusions. It is not so much that each of these disciplines is concerned with separate phenomena (although to certain extent they are!) as that they differ with the methods by which to study perception. If factors of cultural differences – such as history, beliefs, social groupings, patterns of life and work, geographical factors, weather, topography, agricultural land

Figure 5.1 A 'Good Man's Land', near Bendochy, Perthshire, Scotland. Photo: Mary
         Modeen.

use, flora and fauna – are added as variations in *what* is read, then the resultant complex combination of perceptions, interpretations and languages to
convey these understandings expand exponentially.

This chapter will consider aspects of how it is that we humans know what
we think we know, how we variously perceive the world, how we remember
experience and how cultural values affect our memories. Writing as an artist
who has her own understandings that have come through artistic practice,
augmented by an academic education and career as a researcher, it seems
apparent that what we humans see and don't see is fascinating and compels
reflection. Invariably, the language of an artist/academic will have a different sound and contribution to this discussion as opposed to that of, say, a
cognitive neuroscientist. While there is much to gain from that field, and
many others whose focus impinges on the nature of sensation and perception, or artistic representation and hermeneutics, the language and focus
differs of these others from artists, but their work overlaps areas of shared
concerns. In a multidisciplinary investigation, there are many facets of this
topic that reflect each other's research, despite differences in methods and
idiom. In drawing these threads together, it is the intention here to look precisely at the synergistic effect of perception through the unavoidable multitude of cultural inflections.

As a first step, it is useful to review briefly how we humans take in the world around us. *Perception* is not solely the physiological activity of the eyes, nervous system and cerebral cortex. Rather, it is a rapid and complex combination of sensory stimuli and multi-faceted interpretation; the latter constitutes aspects of subject-specific experience, education and expectation, of whole-body sensation and of culturally influenced values and philosophical beliefs.[3] It is precisely this set of phenomenological filters that shapes the interpretation of the neurological impulses received in the brain, and the assignment of meaning. As adults, we see largely what we expect to see.

From the outset, it is clear that differences in culture are inherent in how the world is perceived: the learned influences, as cultural shapers, are profound and pervasive. But a culture that is not respected, or which stands in opposition to the hegemony of mainstream and majority society, loses by degrees its ability to offer its own names, then its own language, and finally, its own values. Here in this book, for example, having to rely on the English word 'spirit' to carry these multiple meanings of 'presence, totem, deity, immortal, ancestor, non-corporeal being' as a translated word from indigenous belief systems (Māori, Ojibwe, Sámi and so on) itself demonstrates language's ability to limit or shape thinking in the minds of the listeners or readers. I will return later to this notion of language's shaping power.

Beyond this intricate subject-specific combination of factors, lie those 'other sensors' about which Anglo-European academics feel uncertain. These non-metric receptors have no English name, no discreet acceptance by European-tradition scholars other than as pseudo-metaphysical phenomena. As Māori peoples refer to *wairua*, for example, or Ojibwe examples include *totems* and *manitou*, various Indigenous peoples especially, but shamans, poets, artists, some metaphysical philosophers and others have sensed and acknowledged aspects of experiential knowledge that are not within the spectrum of most European-based academic traditions. Rilke, for example, in his *Duino Elegies*, lays out a cry to the universe in these mystical, lyrical poems, addressing his thoughts on suffering and anxiety, aspiration and uplifting paroxysms of euphoria and spiritual awakening, as well as his cries of despair at the inevitability of death.[4] We must add here in the fullness of this discussion, that metaphysics and metaphysical poetry and art – especially in an area closely aligned to this subject of sensing the numinous – does indeed seek to explore phenomena as they are apprehended by human consciousness as a *synthesis* of elements that are logical, physical, ontological and spiritual. In short, they merge the numinous with the phenomenal, seeking to promote and reveal an 'inner' sense, emergent in the event or lived experience.

This is where the value of storytellers becomes universal. Not only storytellers, of course, but novelists, poets, artists, balladeers, playwrights and performers who enact narrative creations and who offer narratives that frame experiential perceptions. The arts are built upon an implicit understanding of the value of the imagination, of the truth that underlies singular experiences, and in the importance of hearing 'deep' messages contained within the story. One woman's story is the story we all share. In an era that is led by scientific enquiry this basic premise upon which the arts and humanities are conducted may be overlooked, or devalued. There is an implicit assumption in studying the arts, that there are enduring values and important insights to be gained in listening to the stories of individuals. Elsewhere in this book, discussions of vibrancy point to the ways in which energies reverberate and resonate with contact and exposure to vibrant actants. In parallel with this, it is timely to reinforce that narratives have a similar potency and potential to affect shared insights, and note that they ultimately may lead to a shift in perceptions of the relations of an individual to her environment.

Lucy Lippard understood this, as she wrote in her book *The Lure of the Local*.[5] In this work even the structural layout of the text reflected her underlying intentions; across the top is a running first-person account of her experiences living and travelling in several places across the USA. Underneath is a more historical and critical account of how various methods and disciplines approach the understanding, history and value of place. The overlay of the different textual approaches to place are physical juxtapositions that play out her theories of many years earlier, in *Overlay*.[6] Valuing 'the local' in her terms means embracing as much of the human history as possible to excavate and recover, effectively to augment the fuller narrative of place. This includes looking at conflicting accounts, individual experiences, traces of labours and activities, and setting aside as much as possible prejudices about 'who' in an effort to collect the 'what'. And finally, in this experimental book design, Lippard speculates in extended captions for each illustration what each artwork or image begins to suggest in light of a 'local' significance.

This particular example of Lippard's work from an earlier generation typifies many of the factors highlighted in this chapter. First, the insertions of the autobiographical details in the top section are not by chance. She values the importance of first-person accounts in the manner of seeing the universal in the singular. Orality and oral culture clearly have an important place in her book, offering a model for shared 'local' knowledge. Inserting her own stories into the discourse she offers an implicit comparison of the anecdotal with the historical, the over-inscription of the individual with social awareness. And second, the inclusion of photographs with extended captions is not so much 'documentation' – which implies a singular means of proving one indisputable fact – but rather, an additional perspective, visual material, to stand alongside the other two types of 'conveying local knowledge'.[7]

As individuals, one of the first things we learn is 'our place' in this world. (This is actually an ongoing process throughout our lives, and mentioned several times throughout this book.) We come to know our bodies as infants. We learn what-is-not-our-bodies, by tasting flowers, touching worms (and unfortunately sometimes the reverse) and so on, as experiential knowledge primarily acquired through the senses. Then we are taught the names

Figure 5.2 The Butoh dancer, Nonoko Sato from Kyoto, and musician, Jerry Gordon, are here performing a Butoh piece. The performance took place at MIIT House, Osaka, Japan. Butoh is a form of performance/dance which emerged in the 1950s in Japan, which reacts to over-formalised theatre by embracing the grotesque through crude gestures and the evocation of psychic pain. The enacted negotiation between theatrical space and the space of the imagination is an appeal to empathetic understanding. Photo: Brian Allen.

for these things: flower, worm, *baum, ciel.* We begin to understand differential and relational contrasts: 'Here' is this place that I am standing versus 'There', the place beyond my reach. As adolescents or young adults, we extend this experience to temporal concepts: mortality is the ultimate map, defining the parameters not just of geographic place but of existential place (Figure 5.2).

* * *

Let us take a brief but informative interlude for a wild leap from Kant to the 20th century. We are reminded by the 20th-century Continental phenomenologist philosophers of the insistence on the act of perception. This is a type of epistemology: a study of how we know what we know. This concentration on perception as a product of the individual's senses is opposed to what Kant, in the late 18th century, called things-in-themselves (*Ding-an-sich*). According to Kant, these *Dinge-an sich* are objectified and idealised understandings of the world, the *noumena*, and knowable only through the mind as opposed to the senses. (This is in contrast to the *phenomena*, the thing or event that we know through the senses, as they appear to consciousness; in phenomenology it is always *someone's* experience, with an emphasis on the first-person aspect of knowing.) Phenomenologists and the cognitive scientists of today would argue that embodied perception – the *lived experience*-is the first way we truly know the world in the sense that our bodily senses inform of us of what surrounds us. Our senses are the way by which we physically take in the stimuli that surround us, transmitting this information neurologically to the brain. Contemporary post-structuralist philosophers would say that it is the fact that our own 'always already' being is actually double: in other words, first, as our bodies as both object (*corpus*-that which contains our sensory apparatus) and subject (the conscious embodied-being who feels ourselves being touched).[8] We can touch our own hand with the other, sensing the touch while simultaneously feeling the touch of the other hand. And second, because we cannot think back to a time when this was not the case, the simultaneity of sensing the world is being aware of being the Sensor. Simple. So why is this important?

There is significance in this because this dual aspect of knowing what we know (as lived experience) allows us as both sentient and cognisant humans to understand other humans on the basis of our own experience. This *empathy* – or the empathetic imagination – allows you to know your own subjectivity – through your own bodily experience – and by extension, to understand the subjectivity of others. You can understand the pain of Others, the emotions, the intentions and so on. This is referred to as *intersubjectivity*; it is by this means that one can understand one's own self as an Other to other people. It is a synchronous knowing from the inside and the outside. It creates a space for knowing from multiple perspectives by understanding the states of being by being both Seer and the Seen. And not just seen by a subjective viewer, but Seen by many viewers/Seers, each with his/her own various subjectivities.[9]

The one note to add in this consideration of *empathetic imagination* is that we must remember that while we can empathise with what another human senses, and speculate about how they may understand their experience, because we have not had their experiences or been in their place, it is always a speculative knowledge. We cannot truly know how they react, nor must we assume that all people think or feel as we feel. Their perceptions are unique to their own cultural and personal circumstances, and the range of individual characteristics is part of the diversity that must be preserved. An example of this may be found in the use of the word 'community', a frequent term that has the potential to stand for a whole group of people, whether they are grouped by living in proximity to each other, or by sharing a faith, or a gender, or planting allotment gardens. To treat this as a homogenous 'whole' is to fall into the trap of *essentialising*, of pre-judging on the basis of making a collective unit replace distinctively different individuals (Figure 5.3).

* * *

Figure 5.3 Professor Jiao Xingtao and student artists, collaborating on a project in Yangdeng Town, a rural community on the Yangdeng River, in Tongzi county, Guizhou province, China. Over a five-year period, this extended project combined artists and villagers in many creative collaborations, seeking to re-establish art as a core part of daily life. Here, Chinese sculptor Hong Yang works with local carpenters in Yangdeng village. Photo: Jiao Xingtao.

End of the interlude. Let us return to the question at hand. What has this told us? This knowledge of multiple perspectives that each of us carries within us, then, lays the foundation to allow us to understand what we know of the world from multiple points of view. For instance, we are animal beings, sensate and self-aware, knowing the ground beneath our feet; the horizon line arranges itself at exactly our own eye level. Beyond that, we are also *homo ludens*, capable of play, of abstract thinking and speculation, of imagination and of forward thinking.[10] We can imagine the landscape from a viewpoint of flying even when we cannot bodily take flight. In our imagination, we see the hills, trees and towns from high above; many of us dream in visions like this. Speculative imagination is another way to describe this ability to derive from embodied knowledge in order to deploy that understanding in envisioning alternative environments. Creative practitioners engage in this deep power to imagine the unseen when they see what Le Guin called 'the larger realities'.[11] We know, too, the illusions with which we contrive to trick ourselves – the oases in the desert, and the mirages of youthful selves that stare back at us from the mirror. The appearances of the world around us are mediated by our eyes and our brains; we know this in a rational sense, of course. When we stop to think about this, we know that we are *always* situated, *always* contextual: this is *Dasein*, literally *being there*, equally weighting the state of existence with place, a being-in-place. We are not just the existential objects of what Kant calls things-as-they-are (to others; 'phenomena', in other words); we are always in states of being, being-here-and-now in constant flux, relational to the environment.[12] And then this constantly inflected (and affected) state is further mediated by our perceptions, skewed by various elements such as past experience, expectations, cultural associations, education and so on. The past profoundly shapes our ability to perceive; what we perceive shapes what we will see, now and in the future.

This last statement foregrounds the importance of memory. Memory itself is an elusive concept; it may be described in physio-biochemical terms as a process in which memory is 'laid down' as deposits of ribonucleic acid (RNA) and various proteins in neuroanatomy. This occurs specifically, as cognitive neuroscientists tell us, in the hippocampus and prefrontal cortex of the brain. As three separate stages in the functional process of memory, scientists list encoding, storage and retrieval. This is a useful model for our discussion.

First, 'encoding' may be considered as the first stage in producing memory, speaking from a non-scientific standpoint, more as an artist than as a cognitive psychologist. Encoding is a twofold process. It is the way we make sense of material that we take in; it transforms chaos into manageable bits of order. For example, we might see a telephone number in front of us. 'Chunking' the number into pairs or triplets of numbers is a way to group together digits in manageable size. One way to know this happens is to say your own telephone number: you probably have a rhythm or grouping for reciting these from memory. If someone repeats this number in a different grouping system, your own number sounds strange, almost unfamiliar. Let us take a more complicated example. You are outdoors, in a place you have never visited before. In front of you, all around you, is a grouping of information that is new: your senses tell you visually which is near, middle and far: these are sorted through the apprehension of atmospheric spatial cues. You see vegetation: sight and smell tell you conifers, heather, bracken and earth. You hear slight buzzing: bees in the heather. You feel warmth and movement on your skin: haptic cues for sun and breeze. Your feet feel slightly uneven: kinaesthetic cues for standing on hilly ground. And so on. The second part of this encoding process is the recognition, or at least the ability to group into patterns of associations. By naming these sensations, grouping and the meaning or significance inferred from this have already occurred first through pre-verbal, and then verbal, association and attribution.

If you could not make sense of the sensations, you would not be able to retain this as a distinctive memory. Uncoded, the landscape would not make sense. Complete confusion can only be recalled as a feeling or impression – indistinct. If you have ever felt wholly disoriented, you will probably understand that a sense of place is a construct in the mind. It is not simply the ground beneath your feet if you are unable to 'make sense' of any other sensory cues. It is a knowledge, a certainty, that you can assemble the continuous flow of various stimuli into a 'reading', a nearly synchronous interpretation, or a construct of space, light, temperature and atmosphere. Visual memory, for example, first depends upon the order one assigns to the optical stimuli that one organises. We remember as well in specific modalities: our visual experience is organised in such a way that we remember as a visual recall, not as language, for example.[13]

The second stage of memory is storage. This is the means whereby the grouped information is 'laid down' in RNA and acids which form layers or coatings on cells in various parts of the brain. Sleep is critical to this period.[14] Deprivation of sleep can lead to the malformation – or even inability – to store memories. For artists, there is a resonant poetic image of the importance of dreaming in this process: literally, we must sleep our past into significance.[15] We cannot remember unless we dream.

The third part of functional memory is retrieval. This is the way that we access the information that is stored. Looking for a document to give the tax collector is a case in point: it is no use having kept the document unless you can find it again when you need it. This is the type of memory that most of us usually refer to when talking about the past: it is bidden recall, available on demand. Of course, there are also unbidden memories, those that come flooding back to us for example when we smell cinnamon and find that a memory of childhood Christmas overwhelms us. Often these are linked to a particular sensation, as in this olfactory example. For an individual to have functional retrieval, there must be a system of cross-referencing as well as the equivalent of 'file labels'; mnemonic devices are various, but include systems which rely on 'placing' the memory somewhere intentional for later return.[16] These, then, are the three stages of memory as a process within an individual.

But let us now overlay the same process as an individual with that of one which is more communal, moving towards collective or cultural memory. Communities encode events and information through various means: town clerk records, village newspapers, radio stations, talk shows, schools, local theatre, documentaries, and last but certainly not least, local gossip. In a manner similar to that in individual humans, communal encoding is a way of sorting, grouping, 'taking in' what is going on and 'making sense' of it. Writing, photos, songs and sound tapes, digital records, lectures, television and film become the substance of archives that are 'laid down' in storage files. And finally, retrieval is the third stage of functional memory, accessing the information; retelling the stories. And stories become histories, preserved narratives, facilitating cultural recall.

Again, as a parallel to the individual, this third stage of the process concerns archives, the formal facilitation of cultural recall. History, records, maps and charts are the substance of retrieval, but what is retrieved is the account of what was 'grouped' in the first place. If the process of 'making sense' of an event is told by one person, then stored, and then later retrieved, all we can know of the event is the singular 'taking in' and the singular narrative retold. While this narrative may be fascinating in conveying one person's story, this is unreliable evidence. Singularity of perspective stands in opposition to the very process of multiple perceptions we discussed above. It might be history, but in historical accounts it is always important to ask *whose* history this is. The events that defy 'making sense' – or the imperfectly formed memories of those to whom the events can't make sense – or even the missing accounts whose memories are not stored (i.e., devalued) are equally history, but may not be retrieved.

These accounts, what we might call 'minority reports', are essential to history and historical processes: first, for the same reason as above, that multiple perceptions are the core of human experience. But also, these recollections are unreliable unless substantiation from other perspectives is offered. This has been the basis of feminist scholarship and Indigenous studies for the last 30 years or more. The Othering of history is the recovery of difference; the reprisal of alternative narrative, more specifically the nuancing of narrative, is the process of valuing multiple perspectives. Quite literally, this is a process to reject singular recall and to replace cultural forgetting with cultural remembering.

Jonker and Till write about aspects of Cape Town's history, discussing 'memorial cartography' as a cross-disciplinary action dependent upon documentary archives.[17] The questions remain, *who* has done the collecting, *what* values underpin the collecting and *how* is the collection indicative of the power that underlies preservation and the authoring of local histories? In other words, it is precisely those who are ostracised by the mainstream – frequently, Indigenous peoples, slaves, criminals, foreigners and the very poor – those who stand to lose the most, who have been excluded from the cultural remembering. It is they who are excluded from the collections and social scrutiny of local history, in part because their versions of the past have been perceived differently, organised differently and construed differently. After all, post-colonial theorists have shown us time and again that 'official' history is 'the narrative of the conqueror'. Silenced narrative, erased accounts and invisibility to subsequent generations are the consequences for communities and cultures who are not valued and who have been subsumed (or silenced) by more powerful forces. Cultural 'forgetting', as Ricoeur reminds us, is the flip side of cultural memorialising.[18]

There can be few more graphic examples of cultural forgetting as applied to place-based culture than the 'disappearing' of the hill in California called *Tahualtapa* (or 'Hill of the Ravens') which was sacred to the Cahuilla people, before it was renamed 'El Cerrito Solo' (the 'Little Lone Mountain') by Spanish missionaries. This version, too, was subsequently changed by Anglo settlers, to become Marble Mountain, at the point when it became a mine. These various name changes, as the artist Lewis DeSoto believes, indicates that "the land has become estranged from itself".[19] And finally, in its most recent designation, it was given the name of the owner of the cement quarry, Mt. Slover, into which it has, in fact, metamorphosed. DeSoto

(Cahuilla/Hispanic), himself of dual Indigenous/Hispanic heritage, has documented this transition of erasure in his photographic project *El Cerrito Solo* (*Tahualtapa Project*) 1983–88.[20] He exhibited these photographs and concrete slabs manufactured from the materials extracted from the current quarry, at the Moderna Museet in Stockholm (1993); they are representations of the sacred mountain that is no longer a mountain, half of which has been quarried and is no longer there at all.

Erasure of a mountain by human mining may have a slower and more natural counterpart in the process of erosion over millennia. In the contemplation and study of 'deep time and place', the act of historicising is a debatable process, made more difficult by the vastness of the time parameters in tackling prehistoric periods in an effort to promote a vision of continuity over change. Ann McGrath and Mary Anne Jebbs do precisely this in their edited anthology entitled *Long History, Deep Time: Deepening Histories of Place*.[21] McGrath addresses this challenge head on, when she ponders:

> how the discipline of history might deal with a chunk of time so voluminous that change itself seems too slow, even imperceptible. The history discipline's expectations regarding the pace of history – of its anticipated speed and slowness may need to change to accommodate this period. We do not necessarily know where we are going. Slow history may take us more deeply inside history, or simultaneously throw us outside history as we know it.[22]

McGrath's point is well made: deep time might connote the entire length of time that *Homo Sapiens* is believed to have existed, and indeed well before, if the conditions of time/place are essential to the understanding of human apperceptions of time itself. Historians might argue that there can be no History (with a capital 'H') without humans, because earlier than that is prehistory – the 'proper work' for archaeologists. Deep time is also a challenge to the scope of this discussion because it demands a certain act of stretching one's imagination about the meaning of the word 'deep'; the earth's ground surface at 60-millenia ago was most likely several kilometres higher than at present.[23] So 'deeper' in that sense, in actuality higher and higher than at present, is closer to the meaning of 'deeper in space'. Various Indigenous Aboriginals of Australia also have complex enmeshed ideas of deep time/place; they often interlink geological formations, entangled narratives that are personal, collective, remembered and omnipresent. For them as for us, 'time is multi-layered and mutable'.

> In order to tell the story of a peopled landscape story of long duration, diverse kinds of research teams, forms of evidence collection, narration and analysis are required. If historians are interested in joining such teams, they will need to develop a different orientation, new training, and a change of gear.[24]

In thinking through what is to be gained by contemplating deep time and place, and bearing in mind the profoundly connected but non-linear narratives and experiences of the many Aboriginal peoples whose cultures are closely tied to places with metaphoric and physical pasts, perhaps the very reliance on multiplicity as a framework for combining such a complexly infolded set of stories, legends, creatures, sites, songs, walks, marine navigations and family identities can help to suggest for more Western-based peoples a rich alternative to the academic, philosophically empirical system of specificity, singularity and evidence so favoured in contemporary culture. And if the perils of forgetting are brought to the fore as a fear linked to this type of inter-connected knowledge, one has only to look for examples of Indigenous knowledge that offer complexly based practical knowledge and skills underpinned by intimate understanding of vast networks of entire ecosystems to see that this terror is biased and unfounded.

In contemporary engagements with place in creative, imaginative or aspirational modes, one can find a great many experiments in alternatives to globalisation, to capitalistic economies and models of habitations by owners/farmers/stewards. For example, the island which housed a copper mine in the Seto Inland Sea has become an art complex.[25] Initiated by Yanagi Yukinori, with the collaboration of a wealthy patron, Fukutake Sōiichirō and of an innovative environmental architect, Hiroshi Sambuichi – the enormous Inujima Project, is called *Seirensho*. The former mine site houses a collection of memorabilia that belonged to the national literary giant, Yukio Mishima, as well as including other "contributions to the site at Inujima, notably the renowned architect, Kazuyo Sejima and the leading Japanese contemporary art curator, Yūko Hasegawa, who has acted as artistic director of the island's art house projects since 2010".[26] As a site, a re-invented purpose, a socially engaged cultural contribution, and as an end in itself, this is yet another example of how the arts may cross the divide between memorialising and preserving, initiating and rethinking, the various values inherent in geo-locational sited culture.

Another sited large-scale artistic project that has been undertaken over the last 20 years is *The Land* project in northern Thailand (Figure 5.4).

In writing about this project, initiated by Thai artists Rirkrit Tiravanija and Kamin Lertchaiprasert, and then contributed to by a great many other artists, architects, designers (Superflex), local farmers and others, a sustainable community project for on-going (periodic) artist residencies and volunteer rice farmers and gardeners, fish biologists, solar engineers and labourers has been extant for over 20 years. During that time, innovations and living experiments in working the land have coincided hand in hand with designs for capturing biomass gas emissions, solar energy, simplifications of architectural structures with materials appropriate to the setting and land drainage have all been tested. But in addition, Rirkrit's intention to merge his artistic ambitions with sustainable living and communal collaboration has led to other spin-offs, often designed and built for large Biennials or

Figure 5.4 Rirkrit's House, at *The Land* project, with detail of one of the accommo-
       dation structures, Rirkrit Tiravanija and Kamin Lertchaiprasert, 1998–
       present. Photo: Rirkrit Tiravanija.

Art Festivals,[27] and then these works are disassembled and reconstructed
on the Thai site near Chiang Mai. But this project is not without its critics.
Distanced away from the mainstream artistic world, ironically it requires
significant travel (and increased carbon footprint) by the artists and eco-
art-tourists, as well as Rirkrit himself who resides in New York, to come for
periodic visits. Also, as Rirkrit stresses in his website for *The Land Foundation*, this project is 'not about art', but is about providing an 'open space'
more interested in 'sustainable infrastructure than outdoor sculpture'.
However, in frustrating pragmatic reality, the property has had to be fenced
to keep the plants and fruit trees from being pillaged by local inhabitants.[28]
This was in direct contrast to the origins of the project, with its intention
to provide sustainably produced food for its residents and for HIV-afflicted
local people, which it continues to do.
    In writing about this project, Janet Kraynak said,

> . . . *The Land* can be set apart from the materialist practice of critical
> negation, as well as from the tradition of artistic activism that sought
> to replace objects, however loosely defined, with the orchestration of
> situations to bring about social change. In contrast, temporary meta-
> phors of openness are invoked as frequently as ones of spatial distance,
> in order to unhinge the project from any such requisite outcomes – as

well as to accommodate the nomadic nature of its occupants, structures and activities.[29]

She continues to say that: "*The Land* functions not simply as a place but as an idea, or, in other terms, as a brand that provides an imprimatur on and lends coherence to a diverse, scattered array of activities".[30]

Clare Veal,[31] commenting about Kraynak's critical stance, responded:

> . . . Kraynak fails to address how *The Land*'s discursive, and ultimately fictional forms may impact readings of the work, or how these forms may be consciously constructed dimensions of the work itself. Furthermore, given that Kraynak herself had not visited the site, her article itself is a pertinent example of how the work is experienced in its fictionalised form.[32]

The point to make in this chapter dedicated to polyvalent perception and cultural memories of place, is to examine the critical reception to *The Land* project and specifically to note that there is a separation between the physical location and sited structures and, by contrast, the site as *praxis*. It is (at least) two separate things – the *actual* land and physical structures, and the *idea* of this. Thus, a space opens out between the two aspects of the work. The rice paddies exist, the built structures exist and change over time,[33] the pond and the biomass gas capture unit. But for many, the *idea* of the project as a concept is one that moved away from a model of ownership, one which was intended for cultivation as 'an open space', towards 'discussions and experimentations in other fields of thought'.[34] The polyvalent aspect was thus its artistic origin, with architectural structures by Kamin and repurposed sustainable land use by Rirkrit. As a comparative example, these are in direct contrast to the neighbouring Thai homesteads and rice fields. The aspect that plays out as cultural memory is as much cultural forgetting: in over-inscribing new technologies, and Western (nomadic) patterns of seasonal visitations by international artists and students, teaching projects, meditations and alternative plantings, the multiplicity of new patterns of lifestyles, cultures and interpretations stand in direct contrast to the traditional local Lanna customs.[35] But its contrast has also had a positive effect for nearby local rice farmers; after difficulties with increased farming, some of their fields have been given over to other uses now while they grow rice in more controllable fields.

Let us return now to that insistent question of language, and look particularly at how forgetting and remembering play a role here. At the beginning of this chapter, the terms *wairua* and *manitou*, were identified as the names for invisible presences from Māori and Anishinaabek languages, respectively. Of the latter, the Ojibwe First Nation peoples of Canada,[36] there are members of the nation who recall being taught in schools that didn't allow any language but English, in a pattern that has been repeated endlessly across the world. If the children spoke in their native tongue, they were beaten. They were taught that the greatest hope to which they could aspire was to 'pass' for wholly white, to speak like white Americans, and to hide their origins with manners, clothes, haircuts and above all, language.[37] In other words, a process of concealing ancestry and speaking the common language clearly marked the socially designated way to success. The Indigenous language that had its own distinctive capabilities – to identify perceptions that had no equivalent in English, for example – was culturally 'forgotten'. This decreasing population in many (but not all) tribes by marginalising many of the descendants who would have identified as the next generation of tribal members has added to the accelerated decline of traditional cultures and languages. What has been lost is nameless, since the names themselves in the old language – and that which they culturally signified – are both gone or in decline.[38]

These acts of cultural remembering or forgetting are at the core of a discourse around 'spectral traces'; for example, how do we know what we know of a place when we cannot point to it, have no records, and have no names for any lingering sensations? As an artist approaching this question of perception of the unseen, artistic visual statements have the ability to open up repressed histories and cultures, to view places in ways which suggest temporal and perspectival depth, moving beyond singularity and its suggestion of fixed vantage points or singular (colonial?) insistence on only one way of seeing. As we discussed before from a philosophical point of view, fixity opposes the multiple selves we each carry within us, artificially delimiting the multiplicity we experience as both objective and subjective beings. It precludes us from the empathetic inter-subjectivity of which we are all capable by denying the shared humanity (and hence shared knowledge or values) of others.

Moreover, visual art has the ability to appeal to a more profound and less literal resonance that each of us as humans share in our experiences of being on this earth. As sentient creatures we share the knowledge of watching the sky lighten before dawn, for example, or of looking for a horizon that is lost in mist and cloud. For each of us the bodily sense that informs every movement, every glance, is anchored in the centeredness of being in this world. Merleau-Ponty understands the perceiver as both a body that is the means of perception, as well as the self that is the subject of nature, the human subject who has emerged from nature. He states:

> Every perception takes place in an atmosphere of generality, and is presented to us anonymously. I cannot say 'I see the blue of the sky' in the sense in which I say I understood a book, or that I have decided to devote my life to mathematics. . . Every time I experience a sensation, I feel that it concerns not my own being, the one for which I am responsible, and for which I make decisions, but another self, which has already sided with the world, which is already open to certain of its aspects, and

synchronized with them. Between my sensation and myself there stands always the thickness of some *primordial acquisition* which prevents my experience from being clear to itself.[39]

He moves beyond the body as subject and offers—insists—upon a plurality of experience in which all is not within the perceiver's constitution.

It is not so much that visual art *transcends* cultural differences (because it cannot eschew the cultural interpretations with which this discussion began) as rather, that it reaches the possibility of a pre-linguistic way of knowing that is shared by humans. This viewing-as-knowing at (or perhaps just under) the liminal stage of cognisance is the site of visual creative statement, powerful in its ability to move directly, sidestepping verbal mediation. As creatures who have evolved from biological states where instinctive behaviour occurred faster than articulated thought (by virtue of responses that are *reflexes*, mediated by the reflex arc in the neural net pathways of the spinal column), nearly instantaneous patterns of behaviour still indicate the speed of visual thresholds for muscular responses, preceding linguistic responses. Quite literally, we see before we say.[40] This opens the door to visual, non-verbally mediated, statement. Or does it?[41]

We have now considered three aspects in which vision, perception and cognisance merge. The first that was discussed was phenomenological perception distinguished by various philosophers as distinctive between that which can be known through experience and that which can be imagined or known by thought. Second, we discussed memory as a neuroanatomical process conducted by the body in three separate stages, specified as encoding, storage and retrieval. A parallel was suggested between the individual and society in this function. And then, we discussed reflexes and pre-linguistic apperception. This leaves a question yet remaining: are there sensory processes, which may be described as pre-cultural? And if so, is it possible to describe how the organisational sensing and perceptual processes encode or ascribe value in order that memory may have a final 'acultural' shape? Can this shape ever be determined to be 'acultural'?

Instinctively, one might respond that if we assume that language anchors cultural shaping, then pre-verbal perception (that is, seeing before naming) may equate with pre-cultural. But perception is always the action of an individual. The majority of post-infantile humans have language, but with the acquisition of language has come cultural learning as well. Language is never neutral – or as Irigaray writes, "to speak is never neutral".[42] She argues in this eponymous book that the acquisition of language has come laden with cultural values and prejudices. To search for and assign words to an experience is to reach into the conscious and pre-conscious structures that determine speech in a way that may be understood through a combination of linguistics, cultural studies and psychoanalysis. If we accept this premise, we have arrived then at a point of understanding that the processes of memory indicate, at the very first stages of perception – *before they are even encoded* – a basis in

cultural valuing and judgment. To name is to judge. And if we accept Irigaray's premise, then we need to add: To see is to judge.

Here is another way of recounting this significance, in the words of W.J.T.
Mitchell:

> Landscape is a natural scene mediated by culture. It is both represented
> and presented space, both a signifier and a signified, both a frame and
> what a frame contains, both a real place and its simulacrum, both a
> package and the commodity inside the package.[43]

Here, we may consider the few rare and individual cases when a child may
have been reared without language, apart from overt human carers who, in
lieu of any other socialising factors, might stand for 'cultural influences'.
Might these individuals provide a clue to 'acultural' or non-linguistic subjects who could serve as exemplary case studies for this curious phenomenon
of so-called 'feral children'?[44] Unfortunately, not only is the contemplation
of this type of research ethically repellent, but when these rare individual
cases have been located and the unfortunate subjects studied, their cases
have been largely inconclusive, either because the means of acquiring accurate information is largely speculative or else dependent upon the type of
linguistic relation that the subject is incapable of delivering.[45]

It is important to consider here the thoughts of one final group of scholars. Multiple-perspective, multi-cultural readings and interpretations and
non-centred performative viewings of place are the result of visual art – and
especially performative art – that begin with the intention of witnessing in
the fullest sense, of *participating* in the process of 'living purposefully'.[46]
Non-representational theory, that is 'knowledge without contemplation'
as a result of living practice, and the act of daily patterns of living have
been suggested as an alternative model to the focus on iconic representation. Rather than focussing on the moment of heightened significance as
symbolic of a cultural 'tell' – to use a gambling parlance – it is the ordinary
physical actions as they are played out, that culturally define us. As J D
Dewsbury writes,

> attending to that part of the world full of occurrences that have little
> tangible presence in that they are not immediately shared and there-
> fore have to be re-presenced to be communicated. These subsequent
> re-presentations are fraught with difficulties most apparent in their
> seeming inadequacy; problematizing representation is, however, the
> challenge, the solution, towards an engaging reinterpretation of the
> world. The imperceptibles elided by representation include emotions,
> passions and desires, and immaterial matters of spirit, belief and faith –
> all forces that move beyond our familiar, (because) denoted, world.
> These are not light matters for they forge the weight of our meaningful
> relation with the world.[47]

In this attention to what he calls "the folded mix of emotions, desires and intuitions of the aura of places" he underlines the "communication of things and spaces and the spirit of events. Such folds leave traces of presence that map out a world we have come to know without thinking".[48] These Deleuzean folds to which he refers are intensifications of the exterior world as 'folded in' to the human subject.[49] In this paradigm, culture (i.e., effectively, the external world as shared with co-inhabitants) becomes part of the individual, folded into the fabric of the subject. This is no longer a binary opposition between nature and culture, nor does it allow of simple opposition: rather, it has been completely rephrased as culture *in* nature, embedded (folded) first in the act of perception and latterly in memory.

Here is also what we hear in these words: is this not yet another voice claiming that we must concentrate on the "occurrences . . . with little tangible presence", as Dewsbury says, these things that are "not light matters"? It is by focussing on these things of importance – the nuances of everyday life as it is lived – that we find, value and honour difference, that, while we acknowledge how we perceive these phenomena through the folds of exteriority turned inward, we experience them nonetheless. They are what gives meaning to our lives, 'leaving traces', accruing multiple layers and mapping out the territories of our existence, grounding us complexly in the very place our temporal existence resides.

Notes

1 See for example *The Booke of the Universall Kirk of Scotland: Acts and Proceedings of the General Assemblies of the Kirk of Scotland from the Year MDLX*, ed. by Church of Scotland General Assembly (Edinburgh, 1845), 834. '. . . Anent the horrible superstitioun used in Garioch and diverse parts of the countrey, in not labouring ane parcel of ground dedicate to the Devill, under the name of Goodman's Craft. . .'

2 This appellation gives an entirely different slant on Nathaniel Hawthorne's story 'Goodman Brown' (1835).

3 Notably the work of Maurice Merleau-Ponty, *Phenomenology of Perception*, trans. by Colin Smith (London: Routledge and Kegan Paul, 1962) trans. rev. by Forrest Williams (1981; repr. 2002). The study of sensation and perception constitute an entire field of psychology and could not be adequately summarised here.

4 Rainer Maria Rilke, *The Duino Elegies* trans. by Vita Sackville-West and Edward Sackville-West (London: Hogarth Press, 1931).

5 Lucy Lippard, *The Lure of the Local: Senses of Place in a Multicentered Society* (New York: New Press, 1998).

6 Lucy Lippard, *Overlay: Contemporary Art and the Art of Prehistory* (New York: Pantheon Books, 1983).

7 This type of additional visual material is echoed in the works of W. G. Sebald, notably in his novel *Austerlitz*, where uncaptioned photos purport on one level to evidence textual material, but on another level stand entirely apart and show their own story. W. G. Sebald, *Austerlitz*, trans. by Anthea Bell (New York: Random House, 2001).

8 Paul Ricoeur used this phrase first in his *Time and Narrative*, however, it is also relevant to Heidegger's construct of *Dasein*—literally 'being there' or being in

place—which anticipates a state of being 'ahead of itself'. Paul Ricoeur, *Time and Narrative* (*Temps et Récit*), 3 vols, trans. by Kathleen McLaughlin and David Pellauer (Chicago, IL: University of Chicago Press, 1983–85), 57; Martin Heidegger, *Being and Time* (Oxford: Blackwell, 1962).

9 Intersubjectivity may be defined as 'shared subjectivity between two or more people'. Here, the term 'intersubjectivity' is used in the philosophical rather than psychological sense of empathy.

10 Johan Huizinga, *Homo Ludens* (Boston, MA: Beacon Hill Press, 1955).

11 Ursula Le Guin wrote:

> Hard times are coming, when we'll be wanting the voices of writers who can see alternatives to how we live now, can see through our fear-stricken society and its obsessive technologies to other ways of being, and even imagine real grounds for hope. We'll need writers who can remember freedom—poets, visionaries—realists of a larger reality.

Ursula K. Le Guin, 'Freedom', in *Words Are My Matter: Writings about Life and Books, 2000–2016*, ed. by Ursula K. Le Guin (Easthampton, MA: Small Beer Press, 2016), 113.

12 Again, *Dasein* is temporally specific (now) as well as locational (here).

13 Modes of memory are discussed for example by Spolsky: 'the brain takes in information about its environment through several channels simultaneously, each specialised for the reception of a particular kind of perceptual data. We hear, see and feel, for example, through different systems, and we apparently also store information according to the modality of its reception'. Ellen Spolsky, *Gaps in Nature: Literary Interpretation and the Modular Mind* (Albany: State University of New York Press, 1993), 4–5.

14 Honor Whiteman, 'How Does Lack of Sleep Impair Memory Foundation?' *Medical News Today* (10 April 2017), <http://www.medicalnewstoday.com/articles/316863.php> [accessed 10 April 2020].

15 In a wonderfully similar comparison, many of the Aboriginals believe that the future must be dreamed by their elders: there is no future unless it is dreamed. 'Aboriginals believe in two forms of time; two parallel streams of activity. One is the daily objective activity, the other is an infinite spiritual cycle called the 'dreamtime', more real than reality itself. Whatever happens in the dreamtime establishes the values, symbols and laws of Aboriginal society. It was believed that some people of unusual spiritual powers had contact with the dreamtime'. Fred Alan Wolf, 'The Dreamtime', in *The Dreaming Universe: A Mind-Expanding Journey into the Realm Where Psyche and Physics Meet*, ed. by Fred Alan Wolf (New York: Simon & Schuster, 1994).

16 'Cross-referencing' here, refers to cross-modal memory. Smell is one mode of sensory stimuli taken in, and then stored in memory. For further discussion of modal memories, see Richard J. Allen, Graham J. Hitch and Alan Baddeley, 'Cross-modal Binding and Working Memory', *Visual Cognition*, 17, no. 1–2 (2009): 83–102, <http://www.tandfonline.com/doi/abs/10.1080/13506280802281386>.

17 Julian Jonker and Karen E. Till, 'Mapping and Excavating Spectral Traces in Post Apartheid Cape Town', *Memory Studies*, 2, no. 3 (2009), 303–335.

18 Paul Ricoeur, *Memory, History, Forgetting*, trans. by Kathleen Blamey and David Pellauer (Chicago, IL: University of Chicago Press, 2004).

19 Anya Montiel, 'Reclaiming the Landscape: The Art of Lewis deSoto', *American Indian*, 13 (2012), 24–30. Montiel is Tohono O'odham/Mexican.

20 Lucy Lippard, 'Place and Histories: Writing Other People's Memories', in *The Intelligence of Place: Topographies and Poetics*, ed. by Jeff Malpas (London: Bloomsbury, 2015), 56.

21 Ann McGrath and Mary Anne Jebbs, eds, *Long History, Deep Time: Deepening Histories of Place* (Canberra: Australian National University Press, 2015).

22 McGrath and Jebbs, *Long History, Deep Time*, 2.

23 Ibid., 4.

24 Ibid., 3. A further embedded reference in this quote refers to S.G. Haberle, and Bruno David, eds, 'Peopled Landscapes: Archaeological and Biogeographic Approaches to Landscapes', *Terra Australis*, 34 (Canberra: ANU E Press, 2012), 472. Recent advances in the field including Robin, Libby, 'Histories for Changing Times: Entering the Anthropocene?', *Australian Historical Studies*, 44, no. 3 (2013), 329–340; and Geoffrey Blainey, *The Story of Australia's People: The Rise and Fall of Ancient Australia* (Melbourne: Penguin, 2015).

25 Adrian Favell, 'Socially Engaged Art in Japan: Mapping the Pioneers', *Field*, 7 (2015), <u><http://field-journal.com/issue-7/socially-engaged-art-in-japan-mapping-the-pioneers></u> [accessed 10 April 2020].

26 Favell, 'Socially Engaged Art in Japan'.

27 Such as the São Paulo Biennial, Brazil (2006) and Whitney Biennial (2006).

28 Rirkrit Tiravanija, 'Night School', seminar presented at New Museum, New York (25 September 2010).

29 Janet Kraynak, 'The Land and the Economics of Sustainability', *Art Journal*, 69, no. 4 (2010), 23.

30 Kraynak, 'The Land and the Economics of Sustainability', 24.

31 Clare Veal, 'Bringing the Land Foundation Back to Earth: A New Model for the Critical Analysis of Relational Art', *Journal of Aesthetics & Culture*, 6, no. 1 (2014). <https://doi.org/10.3402/jac.v6.23701>.

32 Veal, 'Bringing the Land Foundation Back to Earth'.

33 *Rirkrit's House* (Figure 5.3) was renovated in 2014 to include more tiles and less wood, in an effort to keep out termites.

34 The website for The Land Foundation <https://www.thelandfoundation.org/?About_the_land> [accessed 10 April 2020].

35 'Lan Na, or Lanna' is the term applied to this region of Thailand in the far northwest, meaning 'A kingdom of a million rice fields'.

36 Ojibwe (Ojibway) nation members are called Chippewa Indians in the USA. The differences in the names are sound-based variations of the same Anishinaabe appellation; this tribal nation obviously pre-dated the drawing of national boundaries.

37 The author's grandmother and mother were two of these people.

38 The late artist Susan Hiller, has made several works that touch on cultural forgetting. Among these is a work entitled *The Last Silent Movie* (2007–08), 22-minute audio-visual work, produced as single screen Blu-ray DVD, continuous soundtrack of extinct and endangered languages subtitled on black screens, British Council Collection.

39 Merleau-Ponty, *Phenomenology of Perception*, 215–216.

40 Rene Dubos, *So Human an Animal: How We Are Shaped by Surroundings and Events* (New York: Scribner, 1968).

41 A contentious arena of current investigation in the psychology of perception is the area defined by the overlap of human evolutionary development from a physiological point of view with neurocognition. For example, aspects of *homo sapiens* 'typical neural response patterns may bear traces of earlier stages in evolutionary development. This may be seen in the various speeds of pre-verbal responses, ranging from pupil dilation to autonomic reflexes, like a hand jerked away from a flame. See, for example, Paul R. Ehrlich and Anne H. Ehrlich, *The Dominant Animal: Human Evolution and the Environment* (Washington, DC: Island Press, 2009), chapter 6.] Yet another example of the effects of pre-verbal

subliminal perception is indicated in the range between objective and subjective thresholds, and the apparently different parts of the brain that are stimulated by subliminal fear when visual material is flashed for a very brief period (but just within the objective threshold) as opposed to supraliminal (conscious) fear when visual material is flashed for a longer period. See, for example, Leanne M. Williams and Others, 'Amygdala–Prefrontal Dissociation of Subliminal and Supraliminal Fear', *Human Brain Mapping*, 27, no. 8 (2006), 652–661.

42　Luce Irigarary, *To Speak Is Never Neutral* (London: Routledge, 2002).

43　W. J. T. Mitchell, 'Imperial Landscape', in *Landscape and Power*, ed. by W. J. T. Mitchell (Chicago, IL: University of Chicago Press, 2002), 5.

44　James Luchte, *Of the Feral Children* (London: Createspace, 2012).

45　Called the 'forbidden experiment'.

46　As Henry David Thoreau described in *Walden: Or, Life in the Woods* (Princeton, NJ: Princeton University Press, 2004), originally published 1854.

47　John-David Dewsbury, 'Witnessing Space: "Knowledge without Contemplation"', *Environment and Planning*, 35 (2003), 1907.

48　Dewsbury, 'Witnessing Space'.

49　Deleuze uses an image of the fold to describe the nature of the human subject as the outside folded in, and more complex intermingled through subsequent mergings: an immanently political, social, embedded subject whose nature is complexly stratified through internalisations and externalisations, complexly entwined.

6 Creative communities of practice: towards an ecosophical understanding of collaboration

Introduction

The environmental philosopher Arran E. Gare writes, "if academic work on the environment is ever to become more than the mass production of high-sounding platitudes, it needs to be related to people's immediate situations".[1] This is, of course, equally applicable to art work. Attempting more genuine forms of collaboration, in the sense that we will be using the term here, is fundamental to the various types of work that inform this book. Perhaps one of the most encouraging developments in recent years has been the rethinking of what might constitute collaborative practices between people working in both specialist and 'everyday' fields of experience This is resulting in a change in the way that we understand collaboration itself. If, as the physicist-philosopher Karen Barad claims, "Existence is not an individual affair. Individuals do not pre-exist their interactions; rather, individuals emerge through and as part of their entangled intra-relatedness", then everything we do is ultimately a collaborative act that intra-relates us with the world.[2] Consequently, it could theoretically be argued that the notion of collaboration is becoming so pervasive as to be a redundant term. However, at a pragmatic level, it still raises many questions. It is one thing to read new theories of entanglement; it is quite another to enact those ideas in our daily lives and practices. As long ago as 1995, Arran Gare observed that, increasingly, in education, as elsewhere: "careerists have excluded or marginalized people driven by a quest for understanding", an insight that, as Mary Watkins has suggested, must now be extended even to NGOs and the 'helping professions'.[3] Consequently, the relationship between ideas like Barad's and our actual practice on the ground needs to be kept consciously in mind. The gap between them, and the tensions which that gap opens, is where our real work starts. Unless we address that gap, reading advanced theory will do little or nothing to help us.

What this means here is that, while we must respect the concerns of academic commentators and theorists who wish to 'further enliven' either their own disciplines or the 'intersection of art and activism', we will not adopt the standard academic role of 'offering critical analysis' to suggest models of best practice.[4] Instead, we'll consider issues of collaboration raised by experience, in the mix of muddle and opportunity in which most collaboration is embedded. But this too must be placed in its proper context.

In 1992, the ethnographer A. David Napier, himself an artist-turned-anthropologist and social activist, offered a caustic critique of artists' notions of selfhood that goes to the heart of the relationship or gap identified above. He identifies the typical psycho-social position of the artist as one of "unintended self-sacrifice in the name of the cult of the individual".[5] He castigated individuals who believe that, to thrive as artists, they must deny that they are not in actuality a sovereign self but enmeshed in 'crowds of others', avowing that each of us is: "as leaky a vessel as was ever made", who has:

> spent vast amounts of your life as someone else, as people who died long ago, as people who never lived, as strangers you never met. The usual 'I' we are given has all the tidy containment of the kind of character the realist novel specializes in and none of the porousness of our every waking moment, the loose threads, the strange dreams, the forgettings and misrememberings, the portions of a life lived through others' stories, the incoherence and inconsistency, the pantheon of *dei ex machina* and the companionability of ghosts.[6]

Our ability to recognise that porousness and its implications, not only for our sense of what a self is but also for our assumptions of personal and species exceptionalism, is fundamental to transforming our collaborative practices.

Fortunately, the cult of possessive individualism no longer dominates the practice of art in quite the way it did when Napier wrote his criticism, thanks in part to a greater sense of artists' growing entanglement in forms of social and environmental engagement. However, 'possessive individualism' remains at the heart of the culture of corporate globalism, not least because it is key to maintaining conspicuous consumerism and economies based on continuous growth. This psycho-social orientation, which is part and parcel of modernity, sees creativity, originality and their products as commodities to be owned exclusively by one 'unique' individual or another. Consequently, the dominant culture still takes for granted that identity is a unique personal attribute or possession, the location of which is deep within each of us, a cultural belief enshrined in the figure of the artist.

As Rebecca Solnit suggests, our self is a permeable and distributed set of inter-relationships that include mutual obligations across kin, friends and social networks, alongside a tangle of broader, far-from-tangible, biological, genetic, empathetic, cultural interwoven with a myriad of other factors. To see ourselves as the sole proprietors of a unique and discrete creativity inevitably distances us from a broader sense of community – human and otherwise – and shared values essential to our psycho-social and environmental survival and is, in consequence, incompatible with any commitment to effecting the fundamental changes we need. As should now be clear to the reader, this chapter is not a guide to 'doing' collaboration. Rather it addresses the need to shift our understanding both of what collaboration is and of the 'who' that does it. To illustrate why possessive individualism is such an issue can be demonstrated by a couple of examples.

Possessive individualism and the denial of empathy

Gare's 1995 critique of 'the new international bourgeoisie' includes a quotation from an economist that epitomises the attitude of possessive individualism towards environmental responsibilities. He quotes this economist who said, "Suppose that, as a result of using up all the world's resources, human life did come to an end. So what?"[7] We might assume that, after almost 30 years of social and environmental activism, this kind of attitude would have been modified. However, Jorge Pérez, billionaire, real estate developer, art collector, philanthropist and friend of President Donald Trump, when asked about rising sea levels at a gallery opening in Miami (for the artist Michele Oka Doner, whose work references environmental issues), dismissed the issue. In his view: "in twenty or thirty years, someone is going to find a solution . . . besides, by that time, I'll be dead, so what does it matter?"[8] In both these cases, an individual cannot see, or perhaps more accurately *feel*, anything that does not relate directly to his own, individual, concerns and lifespan, wholly lacking in the empathetic imagination vital to collaboration in any meaningful sense.[9]

Differentiating between forms of collaboration

Art historians, theorists, cultural critics and other experts who comment on the topics addressed in this book often devote considerable time to identifying what they regard as the most advanced or radical thinking and practice.[10] A case in point are those art forms which can be said to serve as 'artistic models' that: "join the aesthetic dimension of experimental and perceptual engagement with . . . commitment to postcolonial ethico-political praxis" in the context of a "sustained attention to how local activities interact with global formations". We have no particular quarrel with this but, as people concerned about unpicking the culture of possessive individualism, we question whether this approach actually helps promote what we call 'transverse collaboration' in practice.[11]

Such 'examples of best practice' tend to focus on identifying 'exceptional', 'advanced', or 'exemplary' collaborations, thus inadvertently promoting certain solutions over others, stultifying further creative imagination, and enhancing the exceptionalism that reinforces the assumptions of possessive individualism. Such exemplary models are also likely to distract us from paying attention to the realities that confront us in the specific places where we find ourselves. This is particularly true if examples are framed solely in a deliberative language that only speaks to the rational, reasoning mind; forgetting that affective language dealing in 'images and stories' is what generates the type of 'emotional responses' on which empathy depends.[12] Images and stories speak to the ambiguous contingencies surrounding our interactions with people and things, and at the time when it is possible for us to make a difference.

One may be intrigued by T. J. Demos' account of Lise Autogena and Joshua Portway's *Black Shoals Stock Market Planetarium* (2001/04), and guess at the complex layers of technological and other collaboration necessary to realise this aesthetic transformation of data across media and models.[13] However, possibly like others among this book's readers, we lack the cultural capital, specialist skills, or range of contacts necessary to carry out such a complex collaborative project. Furthermore, we recognise that most people wishing to engage with issues of place and environment live in circumstances in which they must first concern themselves with paying the rent, with feeding themselves, and so on; or at least to a far greater extent than Autogena and Portway (an established university professor and a professional games designer, respectively). Much as some readers might like to join the artistic elite able to exhibit at Tate Britain, for most of us, our circumstances make this an unrealistic expectation. Furthermore, because of the damaging psycho-social and environmental consequences of the exceptionalism that drives both possessive individualism and the commercial and institutional art world, that ambition may now be both socio-environmentally and psychically counter-productive.[14] All too often, artists assume their own individual exceptionalism, whatever the claims they make for the social value of their work. This fuels the damaging cultural habit of:

> Taking sides in a conflict of opposites in a system of polarized differences of self and other, us and them, subject and object, reason and violence, etc.[15]

. . . a habit that is incompatible with transverse collaboration as a way of creating a more understanding, environmentally responsible, and socially just, world. The feminist thinker Geraldine Finn, in advocating a 'politics of contingency', notes that alternatives to this habitual way of thinking are: "often invisible to traditionally revolutionary strategists". This often extends to people who theorise from what she calls the standpoint of the same 'high altitude thinking' that dominates critical theory in the arts and, indeed, extends to a great deal of the academic writing to which a book like this might be expected to refer. In the politics of contingency that informs transverse collaboration what matters, first and foremost, is 'the specifics of circumstance and context'.[16] For this reason our first example of collaboration comes in the form of a story based on those things.

Stories are catalysts[17]

What follows is a précis of part of Hanien Conradie's *The Voice of Water: Re-sounding a Silenced River*, a story about collaboration presented at the 2018 *Liquidscapes* conference, Dartington, England.[18]

This place is part of myself. . .
My relation to this place is part of myself. . .
If this place is destroyed, something in me is destroyed. . .
My relation to this place is such that if the place is changed, I am
changed.[19]

It was in 2011 during a visit to my sister in The Netherlands, that I discovered the writing of the deep ecologist and ecophilosopher, Arne Naess. As an Afrikaner female, I had no idea where I truly belonged. I realised that even though my ancestors had lived in Africa for over 300 years, I felt neither African, nor European. My body seemed to remember the Netherlands on a cellular level but my heart longed for Africa where I was born.

The opening quotation consists of excerpts of Arne Naess's thinking about place. He concludes that because of our *relation* to a place, the destruction of an environment is a threat to our inner selves. Naess encourages us to go deep into ourselves, our local places and nature, so we can trace the roots of ecological crises in our own contexts; he says this would assist in understanding global environmental contexts.

I decided to experiment with Naess's suggestion and started to make a list of places I felt I already had some relation to. One of these places is a small farm in the Breede River Valley, where my maternal grandmother farmed with export grapes in the 1950s. My grandmother named the farm 'Raaswater', after the raging sound of the Hartebees River that coursed through the property. The Hartebees River is one of the many tributaries originating in the surrounding mountain ranges. These rivers culminate in the Breede River that meanders gently through the centre of the valley.

The river on Raaswater formed a large part of my mother's childhood memories. Her memories became mythological stories which taught us her understanding of life. The continuous sound of its rushing waters was a reassuring presence which formed the backdrop of every moment lived on the farm.

Raaswater became an idyllic place in my imagination. I yearned for being part of the peaceful symbiotic existence between the natural environment and humans, as suggested in my mother's stories. By the time my mother was married, Raaswater had been sold and so we never had the opportunity to know it.

A quick internet search revealed that it was currently run as a guest farm. I called the owners, explained that it was my mother's childhood home and requested permission to visit. And so a week later my mother and I set out with boundless expectations to go and re-live her childhood memories in this extraordinary place.

The approach to Raaswater was ominous. After we passed a highway petrol station we drove between a densely populated suburb and dry desolate koppies adjacent to the 'Karoo Desert National Botanical Garden'.[20] My mother became disoriented and we missed the turn off to the farm. Eventually we found our way, crossed a bridge and turned into the electric gates.

As one would expect, we found the old thatched-roofed farmhouse and its garden much changed. We walked down to the river to spend time with a part of the farm which we presumed would have remained much the same.

The approach to the river was through a fence and past an organic recycling dump in a field covered in fine grasses and very little else. It seemed like a landscape that had been ploughed too many times and only a limited variety of introduced plant species remained in the soil. I could not see any recognisable indigenous vegetation. As we pushed through some thick alien tree species, the river came into view.

What greeted us was a silent river. Dry river stones and a few puddles of polluted water lay muted in a sulphur stench. A thick black telephone cable looped high across the river and continued over the barbed wire fence demarking the boundary of the property. In one of the still ponds floated the remnants of a child's game: a plastic Barbie doll missing its arms and legs and a similarly dismembered baby doll. My mother had become as voiceless as the river and was slowly tracing its banks. I deeply regretted bringing her on this journey.

The visit to the farm threw my mother into a depression that was palpable for weeks. My guilt over her state of grief was amplified by a realisation that my ancestors were unknowingly complicit in the destruction of the indigenous features of the landscape. How do I connect with a place that is bereft of any indigenous life; a place that is now completely manipulated and changed by humans? How will I ever know what the landscape's true nature is?

After this visit I spent a long time interacting with the people of the Breede River Valley. I spoke with farmers, botanists and a local ecologist who helped me to understand what Raaswater would've looked like in its indigenous state. The river would have been flanked by a now endangered vegetation type, namely the Breede Alluvium Fynbos, a feature which forms part of this vegetation community is a seasonal wetland, which controls the groundwater of the ecosystem; in summer the soil dries out to a certain extent, whilst in winter the ground surface is covered by water.

My challenge was to find ways to connect to a place where the main factor was loss. The pliable ochre-yellow clay of the riverbed is effectively the only living native aspect of the landscape that remains, and is central to my mother's childhood memories about the place. I collected some clay and took it to the studio to see what it would reveal to me'.

After much time of relating to the clay (watching it crack as it dries, rolling it over surfaces to create liquid landscapes, filming it as its wetness dried) I realised that with the help of water and gravity, the clay made waterfall-like surfaces which seemed to elicit a visual soundscape. I had in mind the manifesto of the Gutai Group of Japan, where it is suggested that we do not dominate matter but shake hands with it so it would reveal its characteristics.[21]

It happened to be a full moon lunar eclipse on the night I went to the studio to set up the largest panels I could fit into the room. I covered the panels in a thin layer of clay and spoke a silent prayer that it would create a visual sound, a scape of waterfall-like shapes perhaps. As the eclipse was unfolding I rhythmically and in a meditative state dragged a brush loaded with water across the surfaces to give fluidity to the mud. I went home late that night leaving the dripping panels to finish themselves.

The next morning, I came to the studio and I found that the waterfall effect didn't happen as expected. Initially I was disappointed. It seemed that the clay had moved upwards in capillaries defying gravity. As I contemplated the surface and let my controlling ego out of the room, a new picture began to emerge. I saw a wetland with reeds. I saw Breede Alluvium Fynbos that would have grown on the banks of the river. And finally, I saw that these reed-like shapes resembled an electronic soundscape.

I was astonished. The clay did not behave in the way I thought it would. By relinquishing some control and giving it the space to show itself to me I felt I understood viscerally that the clay had a memory; that it is inextricably part of the ecosystem which formed it and that like described in Jane Bennet's *Vibrant Matter* (2010), this clay is intelligent and alive.[22]

This experience, has changed the way I relate to myself, to human beings and to other-than-human beings ever since. Relating to the living clay connected three generations of Afrikaner women and it allowed me to enter into communion with the plants, people and earth of the Breede River Valley. I started to feel an intense sense of belonging in the world. I listened, asked questions and found *my* place – not really a physical one but a non-physical place of relationship, which somehow anchors me.[23]

Creative communities of practice

The type of transversal ecosophical collaboration that informs Hanien's story and *Rasswater 111* (Figures 6.2) can be understood as work undertaken by a 'creative community of practice' that comes into being through the process of the project itself, providing we understand that 'community' here can include such materialities as river clay!

'Practice' here is understood as a form of transverse action – hence the term 'communities of transverse action' as referred to in Chapter 1. Another important point is that 'community' here is not assumed to be a given or

Figure 6.1 Community groups clear ancient footpaths across the island of Aegina, Greece. Photo: Henri-Paul Coulon. (In Aegina the effort for the locating, clearing, sign posting and promoting the old trails of the island started in 2010 by the members of two local citizens associations, 'The Association of Aegina Active Citizens' and ANAVASI. Nine paths have been completed since, through the collaboration of the local community and the ELLINIKI ETAIRIA – the Society for the Environment and Cultural Heritage as well as the support of the local authorities.)

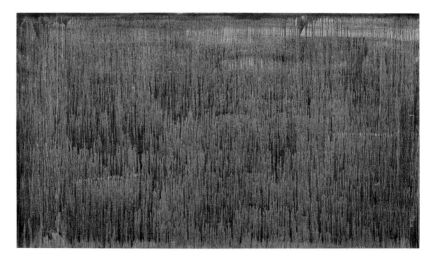

Figure 6.2 Hanien Conradie, *Raaswater III*, river clay on board, 2014. Photo: the artist.

permanent entity, a noun, but is understood as a verb, a 'doing', something brought about by a specific occasion, a temporary trans- and inter- personal constellation of beings, circumstances, materials and skills that cuts across the norms of possessive individualism. It is useful to define some of these terms to clarify what follows.

'Transverse', a term borrowed from Felix Guattari, refers here to a cutting *across, in practice*, of existing social presuppositions, assumptions, boundaries and hierarchies and the various disciplinary, professional, and so of other given, taken-for-granted, social assumptions and structures. The importance of transversal action is that it makes unconventional and unexpected connections without suppressing differences. 'Action' is used here in the sense proposed by the philosopher Hannah Arendt, as that which is vital to keeping-open human horizons of possibility. It refers to our ability to collectively initiate something new and unexpected *together*. This gives 'action' its social value and meaning, specifically as the enactment of mycelial relationships between material environments, social relations, and the inter-subjectivities that animate the ecology of becoming.

These terms relate closely to the question behind this chapter, which is: "What can *we* do to make our collaborations more ecosophically effective"? This needs to be thought through because institutionalised frameworks of all sorts increasingly govern and restrict how people work together, often without them understanding what is at stake. Some understanding of creative communities of practice is inseparable from ecosophical collaboration. Their interdependence lies in the inclusiveness of ecosophy, of the three interwoven dynamic fields of environment, society and the fluid constellation of persona and inter-personal interdependencies that make up a 'self'. Guattari's focus on transverse dynamics, the cross cutting of existing social interactions and flows, is particularly relevant to collaboration. The ecosophical function of transversality that underwrites creative communities of practice is what enables them to address the highly complex, multi-layered eco-social issues we need to face. If an eco-socially viable future is dependent on reconstructing the Commons as a social principle, then *the art of collaboration* becomes an essential social practice.

Collaboration in practice

How does most collaborative work get started, in what conditions, under what kinds of contingencies and constraints?

A project that may be used as an example is one that had two funders with different requirements (one a bank, the other the UK's Arts and Humanities Research Council), and where the personal expectations of the team involved were also different. The team consisted of an architect employed by the bank, a Master's student doing his final year project, a young ceramic artist who had received her first bursary to research innovative uses of paper- clay, and myself.

My first task – set by the university for which I then worked – was to find a way to fold the ceramic artists' funded research into a potential commission to provide artwork for a large bank's new regional headquarters. Work for a bank may seem an odd example to choose in a chapter concerned with ecosophy, but it precisely illustrates just the kinds of complexity and ambiguity that frame so many collaborations at every level.

The bank had indicated what they *didn't* want, and then simply asked us to come up with a proposal.[24] Detailed discussion with the bank's architect resulted in my being able to assemble a team with the official aim of 'enlivening the working environment' (the bank did not want the word 'art' to appear in its annual accounts), while setting ourselves the task of making work that quietly challenged some assumptions about corporate spaces. The team's strategy was twofold: to 'place' the work we made through detailed historical research into the local area, and to give this 'placing' a degree of *social currency* by working closely with three local schools. This enabled us to create a sense of the building as existing both in its own geo-locational past and in a speculative future as imagined by inner-city children.

The process of collaboration largely depended on the team establishing enabling conditions for itself. The architect and I had built a necessary degree of mutual trust and understanding by undertaking a small preliminary project together. Because the team members had different disciplinary skills, backgrounds and needed distinct personal outcomes, we identified a set of shared concerns, values and points of reference early on in the process, which then enabled us to negotiate the inevitable differences that the project threw up as it progressed. This ensured that we could be very clear about what we, as a team, wanted to prioritise and how we would negotiate those priorities both with the bank and with the three schools who became our partners in generating imagery.

In terms of the crucial issue of timing, we were fortunate to be brought into the planning process right from the start. Additionally, the architect, as a sympathetic employee of the bank, was able to mediate our work to his superiors. Finally, we overcame institutional nervousness about aspects of the project both by being flexible (within the limits we set ourselves) and, when necessary, by trying to put ourselves in their place.

The approach outlined above enabled us to fulfil both our own and the bank's aims for the project, and to meet the criteria set by the schools, the Master's course at the university, and the Research Funding Council. However, it is important to note that the final conditions of flexibility and putting ourselves in the bank's shoes, so to speak, hinged on the fact that everyone involved was able and, more importantly, *willing* to be open to whatever contingencies arose, even when this meant moving well outside our individual comfort zones. I will now look at these two points in more detail.

In the first case, it is not always easy for us to understand why a bank's employees think as they do and, in this case, that required temporarily setting aside our own feelings. Fortunately, the differences between the business world and the artwork are not always as great as we would sometimes wish to believe. Both, for example, understand the importance of publicity in building a reputation that will ensure that they have a viable future. The team had worked closely with classes of school children to generate imagery that reflected their vision of Birmingham's future, so we were taken aback when our initial request that they be invited to the opening of the building was turned down. However, through the architect we argued that it would reflect well on the bank if they were seen to have supported an initiative that encouraged children to both explore the city's past and take a creative interest in its future; and that their presence at an opening attended by the regional press would best reflect that support publicly. Fortunately, the bank then agreed.

We were subsequently roundly criticised by some of our more political peers in the art world for working with a bank at all, for providing it with an occasion to mask its 'real social function' (presumably felt by some to be financial exploitation at the expense of all social well-being) behind a display of educational and artistic window-dressing. From the perspective of 'high' theoretical thinking, I understand the logic of this criticism. However, in terms of the lived contingences of the situation itself, I chose to understand the project as enabling a series of educational opportunities: for the team, the children we worked with, and for the bank's employees in relation to the city in which they worked. Whether that choice is justified is for the reader to judge.

In the second case, we had originally agreed to produce paper clay wall reliefs deploying historical imagery, linked to a series of paintings based on children's artwork for the main reception area, along with sets of large and small silkscreen prints based on the same artwork to be dispersed throughout the building. However, as the project progressed it became clear that, if this was to be properly embedded in the larger schema of the building, it would require us to take on additional design tasks not budgeted for in our original negotiations and which none of us, as individuals, felt properly qualified to carry out.

The solution, having negotiated a transfer of funds between budget headings, was for the architect to move from an advisory and mediatory role to work creatively in tandem with the ceramic artist. By pooling their skills, passing over mundane tasks to myself, and seeking advice from specialist suppliers only too pleased to share their expertise, they found ways to use the glass reception desk and glass doors as a basis for etched and transferred imagery. This allowed us to create thematic links to the main work, providing a degree of thematic unity throughout the building and, additionally, create a level of visual signage.

Interdisciplinary collaboration

It is worth continuing to examine in detail some issues that follow from having worked with a small team in a different project, in the second case one comprised of two artists, a web designer and an ethnographer, one of nine teams nested within a major project undertaken by staff and research students from a consortium of British universities. This project, run by social scientists, was predicated on using interdisciplinary methods to explore the connectivity of older people living in rural environments. All the indications are that the funders and most researchers saw the project as successful, a view I do not altogether share.[25]

In the very early stages of the project, it became clear that our sub-project had only been included to meet certain core funding criteria and that the group of senior academics who had assembled the original funding bid had already made most of the key decisions, including the allocation of funds and fine-tuning of aims and practical strategies. Consequently, they assumed that the rationale for their decisions was self-explanatory. Differences about core assumptions, for example that we would all be 'collecting data', were raised in project meetings but were never properly addressed. The geographical spread, size, and multi-centred structure of the project meant that no mediation between majority and minority views was undertaken (Figure 6.3).

Figure 6.3 A recent research project on the island of Ilhabela, Brazil focusses on the traditional knowledge of *Caiçara* fishermen and their families' lives. Pictured here are research assistants Francisco Pereira Da Silva, left, and Patricia Aparecida De Souza, right, interviewing Caiçara fisherman, Osvaldo Clemente. As a collaborative project, scientists, geographers, fishermen, a poet and a filmmaker collaborated on this project (2020). Photo: Mary Modeen.

This example indicates something about the environment in which a great deal of collaboration takes place. The primary environment of any project, ecosophically understood, is its *'taskscape'*, the entanglement of scales, dynamics and temporalities within which the participants work.[26] Institutional approaches to inter-disciplinary collaboration, underpinned as they are by rigid hierarchical and disciplinary assumptions, work against any properly thoughtful engagement with a project's overall taskscape. We have only to think of the range of temporalities involved, from swift, erratic shifts of affect at a project meeting or in a practical workshop to the slow drifts and clashes of public interest that gradually transforms large institutions, to sense why this is the case. A willingness to notice and respond to these fluctuating processes is vital to ecosophical collaboration and needs to be reconciled, where possible, with the rigid world of institutionally driven timetables and protocols. I stress this because an understanding of taskscapes played out in different ecological registers – of subjectivities, socialities and environmentalities – enables us to see beyond academic claims about interdisciplinarity to an ecosophical understanding open, for example, to being traversed by extra-academic and vernacular practices.

Increasingly, most large-scale 'inter-disciplinary collaborations' can more accurately be described as economically or institutionally driven co-operation. They involve people or teams from different disciplines co-operating on the given (and therefore largely tacit) assumptions of professional or economic necessity. Frequently these people, or those who manage and audit their work, have simply calculated that working together gives the group a better chance of getting funding. Consequently, their co-operation as a group is primarily about access to resources and to institutional recognition and reward, and is tacitly structured and delivered accordingly. Economics becomes the dominant context and generates a largely *instrumental co-operation*. If we are involved in working with institutions, we may not be able to avoid some degree of entanglement in this situation. However, we can see instrumental co-operation for what it is and name it as such. To do so publicly is already a step towards more ecosophical forms of collaboration. Instrumental co-operation may pay the rent. It may produce interesting environmental art, enhance professional reputations, and even have positive effects in the physical environment. However, ecosophically speaking it also has profoundly damaging psycho-social implications. Why this is the case has been indicated by Andrew Goffey.

Goffey sees as central to Philippe Pignarre' and Isabelle Stengers' argument in *Capitalist Sorcery: Breaking the Spell* the need to bring into our considerations of power and the political all those issues that "tend to rely on the idea that the public doesn't (need to) think". He adds that this concern "with 'meddling with what doesn't concern you'" can be understood in terms of 'political ecology', requiring us to make visible the politics present whenever an issue is identified as "simply a matter 'for the experts'"; where taken-as-given expert authority becomes just another way of stabilising rou-tine practices and values, of rendering the questionable 'natural'.[27]

If it is increasingly difficult to work in ways not framed by instrumental co-operation, what can we do to intervene ecosophically into this situation? We can start to articulate, and if possible, implement, other versions of collaboration. Collaboration is conventionally defined in terms of working alongside someone (the root of the term is in the Latin *laborare*: to work), with a view to achieving a common goal. It is said to involve a deep, col-lective determination to reach that goal – by sharing knowledge, learning, and, significantly, by *building consensus*. But there are other, less familiar types of collaboration, for example *adversarial collaboration*, as sometimes used in the sciences. This involves work conducted by two or more groups of experimenters with competing hypotheses, and with the aim of both *con-structing* and *implementing* an experimental design or way of working. The aim is to satisfy the differing groups that there are no obvious biases or weaknesses in that design. The significance of adversarial collaborations for ecosophical work is that it starts by openly acknowledging a *lack of consen-sus*, precisely the lack that often lies unacknowledged at the heart of many inter-disciplinary projects based on instrumental co-operation.

Instrumental co-operation, almost by definition, requires people from institutionally 'subordinate' disciplines – for example the arts – to suppress differences in order to 'get the job'. Adversarial collaboration, however, re-quires that people begin by acknowledging that different disciplines work from different presuppositions. Equally importantly, its aim is to find a valid and mutually acceptable way of *doing* something, an effective process. As such it is educational in the best sense because it modifies participants' future ways of forming hypotheses and simply 'working stuff out'.

There is another advantage to adversarial collaboration. It builds in a component of testing and challenge at virtually every stage. When work-ing to achieve a particular action and knowing that team members with very different approaches are already at work on their version, then the care and scrutiny in constructing a carefully designed and articulated pro-posal is more likely to exceed normal expectations among peers who share presuppositions. Understanding this arguably produces an *anticipation* of challenge, and therefore a more thoroughly considered plan of action. It is important to be clear here. We are not suggesting that an ecosophical orien-tation is only possible if collaboration is conceived of as 'adversarial' in this sense. The point, rather, is that, in the spirit of disciplinary agnosticism, we should try to start collaborative projects by identifying and taking seriously the inevitable *lack of consensus* between members of any team. Respectfully done, this offers the possibility of opening-up new and unpredictable, in-deed transversal, ways of knowing and acting. As such it has the potential to induce collaborators to accept shifts in their self-understanding, both as citizens, artists and persons, rather than simply confirming their existing presuppositions (Figure 6.4).

Moving on from these examples, we come to those instances where a community or public become co- creators, collaborating in actions that have as the core intention a larger scale ambition. In these works, the 'socially-engaged' or 'community' artist, does not seek to forge a consensus or to represent one point of view. Rather, they work on a basis of community activation through the arts, devising plans for the inclusion of community groups to affect change through grassroots actions. In these strategies, the artist works with funders or agencies that realise that an entire process is necessary in order to listen to and strategically incorporate the concerns of many people of differing age, abilities, opinions and backgrounds.

One case out of hundreds such around the world is found in Houston, Texas with the Centre for Art and Social Engagement, hosted by the Kathrine G. McGovern College of the Arts at the University of Texas. There, this Centre 'is invested in civic and social engagement initiative that link the College to the surrounding historic Third Ward neighbourhood and the City at large'.[28] Recognising that there are wider implications for projects conducted on a community scale, the Centre has identified three themes that inform its work: the first is Arts and Community Stewardship, the second is The Creative Economy, and the third is Creative Placemaking. Addressing these in reverse order, the Centre has responded to an area already identified by the USA National Endowment for the Arts in its focus on Creative Placemaking. The NEA defines this as 'when artists, arts organisations and community development practitioners integrate arts and culture into community revitalisation work – placing arts at the table with land-use, transportation, economic development, education, housing, infrastructure and public safety strategies'.[29] What is clear here is that the aspects we have just been describing, when artists work in broad collaboration with people from a broad and diverse set of skills, work together, that the beneficial effect is greater than the sum of its parts. (Christine Baeumler has expressed much of this same opinion in describing her practice in Chapter 4.) The artist is often the driving force for recognition of just how creative problem-solving may resolve areas of deficit or dissolution. There is broad general agreement that the arts can be factors for positive changes, often helping the community to see old issues in a new light, and improving people's lives through enjoyable engagement and participation. The co-creative aspect of this work is the key: when colleagues join artists deploying their specialist skills in engineering, construction, urban planning, primary and secondary education, etc. – entire communities can lend a hand in concerted actions. The old adage that 'many hands make light work' is seen in evidence in these transformative instances. Having had a role to play, there is also more of a vested interest in preserving these changes, and protecting the changed areas.

The Center for Art and Social Engagement in Houston identified as its second focus the Creative Economy. In this category, there is ample evidence to see how an invigoration of the arts has the ability to lift the local economy. It happens on a local scale when artists' studios are provided at subsidised low cost by city councils; almost instantly, once the artists move in, the neighbourhood becomes energised: cafes and restaurants spring up, the nature of retail shops changes from low-end retailing to bookstores, music shops, higher-end fashion retailing, and then the other arts become attracted. Dance studios, theatres, nightclubs and community centres emerge.

Designers work with local entrepreneurs to help businesses work more effectively. Architects start designing sustainable and stunning new buildings. Young professionals move back in from the suburbs and reside in the revitalised area where they start families, bringing with them improved schools. This pattern of regeneration has been played out many, many times across the world in urban areas, where the self-organising ability of people attracted by the arts plays out in various repetitions in urban settings.

The first area that the Centre identified is the link between the arts and stewardship, and with all that has been said here, it is evident how this is the result of investment and natural engagement with the arts. When people are included – as actants, as co-creators, and audience, and as influencers – they have already moved more than part of the way toward becoming stewards, guiding, protecting, overseeing. In fact, this transformation from the One to the Many is something that is characteristic of the transformation in collaboration.

Another practice that engages the public is that of Rebecca Krinke, based in Minneapolis. In her public work, the *Mapping of Joy and Pain* (see Figure 6.5), Rebecca made a cartographic replica of a section of the map of Minneapolis/St. Paul, laser-cut and burned into a four-foot by eight-foot section of heavy plywood. She set this up in public parks, and asked passersby to colour in the map on exact spots where, in their personal lives, they experienced memorable joy, and to indicate this in gold coloured pencil. And similarly, they were asked to mark the exact spots where they felt intense pain, and to indicate this in grey pencil. This work achieved several results through collaborative co-creation with the general public. It made site-specific the kinds of personal experiences that are seldom captured in maps; no longer simply informational codes for geo-locational orientation, Rebecca captured and then transformed the map into a deep map, layering emotion, sensation, memory, and personal private experience onto the collective. The One became the Many in this work. As the gold reflects, a great many people found joy along the riverbanks, for example. In one way, this is not surprising, but in another, the affective and emotive sites are transformative; schools, for example, were heavily marked by children who spoke about 'pain, great pain' and also real delight at exactly the same spot. Hospitals, too, were intensely over-inscribed with both colours.

As the title suggests, Rebecca Krinke's artistic strategies were even more subtly nuanced in this public appeal to memory: her title 'Seen/Unseen' leads the viewer to consider once again, that which is not captured on maps, or even in witnessing in everyday life. The pain of others, as Susan Sontag

Figure 6.4 Meg Saligman, *M. L. King Mural: We Wil Not Be Satisfied Until*, mural on the ATT&T building on Martin Luther King Blvd, Chattanooga, Tennessee (2016). Photo: MLS studios. This mural is over 42,000 square feet and one of the largest in the US. Its content was the result of long public consultation and research into the site history and cultural significance.

reminds us, is only truly knowable to those of us who lived through such things, and photographs of suffering, while they have power and importance, must rely on their full narrative and framing, and must never become objects for prurient consumption.[30] Rebecca has avoided this pitfall of close representation and instead, opted to reveal through the code of mapping what cannot be seen – precisely because it is personal, and in the past, that country to which there is no return. Her work grows in its impact as a mute testament to the anonymity of the Collective's joy and pain, human emotions that are real enough but ephemeral and leaving no lasting trace on the places where they were experienced. (Although it should be added that it also served to generate the mutual sharing of personal stories by those working on it, and so a strengthening of a sense of felt community.) In many ways, this work memorialises human life itself, holding on to the unseen shared human experience of what it means to be in this world in all of its states of being, thinking and feeling.

In these instances of transverse communities of creative collaboration, we have traced examples of ecosophical practices, psycho-social and eco-social collaborations, interdisciplinary collaborations where various skills are deployed collectively, and even adversarial collaborations that can be very

Figure 6.5 Rebecca Krinke, *Seen/Unseen: The Mapping of Joy and Pain*, a public
collaboration, Minneapolis/St. Paul, Minnesota, 2010. Photo: Rebecca
Krinke. See the video about its making: www.youtube.com/watch?v=
ZUH6kKP9KsY.

useful. We have also seen examples of collaborations with the public at large
where strategies by individual artists (Krinke), a team of artists (Saligman)
or an entire Centre dedicated to work with communities through the Arts
(CASE in Houston) have led to transformative and experientially influenced
results. In all of these cases, the last consideration here might be that of the
agency of the artist in meeting and working with others. Agency is a word
much used, and it would be typical of most artists to desire some kind of
agency in the aspirations they have for their work. But in the examples given
here, the agency emerges from a single individual who is more than a monolithic entity, in setting into force the transformations that occur when that
individual can devise, create and then co-create with Others, no matter how
they differ, and precisely through the process of valuing differences, in these
communities of transverse action.

Notes

1 Arran E. Gare, *Postmodernism and the Environmental Crisis* (London: Routledge, 1995), 160.

2 Karen Barad, *Meeting the Universe Halfway: Quantum Physics and the Entanglement of Matter and Meaning* (Durham, NC & London: Duke University Press, 2007), ix.

3 Barad, *Meeting the Universe Halfway*, 25.

4 T. J. Demos, *Decolonizing Nature: Contemporary Art and the Politics of Ecology* (Berlin: Sternberg Press, 2016), 11.

5 A. David Napier, *Foreign Bodies: Performance, Art, and Symbolic Anthropology* (Oakland: University of California Press, 1992), 21.

6 Napier, *Foreign Bodies*, 248.

7 Gare, *Postmodernism and the Environmental Crisis*, 12.

8 Jeff Goodell, *The Waters Will Come: Rising Seas, Sinking Cities and the Remaking of the Civilised World* (New York: Little, Brown and Company, 2017); quoted by Meehan Crist, 'Besides I'll Be Dead', *London Review of Books*, 40, no. 4 (22 February 2018), 12.

9 The Marine Reserve extends from Cap Gros, beyond Illa dels Porros and ends at Punta des Morter. Since the Marine Reserve's creation, observers have noted a natural increase of both number and size of the species inhabiting it, it is quite frequent to view red lobsters, groupers, breams, conger eels and many other typical Mediterranean fish specimens here. The richness and diversity of seabirds are also re-established here as a consequence. This protection stands in direct opposition to the political opinions and actions indicated by Jorge Perez, cited here. See: Menorca, 'The North Minorcan Marine Reserve', <http://www.menorca.es/contingut.aspx?IDIOMA=3&idpub=8863> [accessed 10 April 2020].

10 Demos, *Decolonizing Nature*, 12.

11 The term 'transverse collaboration' as used here derives from Felix Guattari. However, as Gary Genosko points out [Genosko, Gary, 'A-Signifying Semiotics', *The Public Journal of Semiotics*, II (I) (January 2008), 11–21], this concept "mutated over time". Born out of a search for a form of therapy adequate to institutional contexts, transversality became identified as the collective property of groups, albeit somewhat confusingly as both "the tool used to open hitherto closed logics and hierarchies", and a process "facilitated by opening and maximising communication between different levels of organisation in an institution" (ibid.). Transverse collaboration is understood here as grounded in "group Eros" (ibid.), as a radical creative collective practice that aims to cut across or rupture existing models of organisation and institution (including those of both the academic and arts worlds), and that gives rise to "subject groups capable of internally generating and directing their own projects", and as simultaneously maintaining both a certain closeness between groups and the larger organisations of which they form a part and an avoidance of any "slide into bureaucratic sclerosis" (ibid.). It is "an adjustable, *real* coefficient, decentred, and non-hierarchical" (ibid.), with a particular emphasis on generating a sense of in-between-ness identified here with the thinking of Geraldine Finn.

12 Rebecca Solnit, *The Faraway Nearby* (London: Granta, 2013), 243.

13 *Black Shoals Stock Market Planetarium* is an on-going installation very like a planetarium installed in the basement of Somerset House, London. It tracks meta-data from the world of financial markets, and converts live capital transactions into lights that seem like shifting constellations in a night sky. Each 'star' represents a trade company, and each flash a stock market transaction. It formed a shifting, metaphoric but ultimately 'un-self aware' aestheticisation of metadata. Joshua Portway and Lise Autogena, *Black Shoals Stock Market Planetarium*, 2000–Present, installation, <http://www.blackshoals.net/> [accessed 10 April 2020].

14 See A. David Napier, *Making Things Better: A Workbook on Ritual, Cultural Values, and Environmental Values* (Oxford: Oxford University Press, 2014), 138–139.

15 Geraldine Finn, *Why Althusser Killed His Wife: Essays on Discourse and Violence* (Amherst, NY: Prometheus Books, 1996), 141.
16 Finn, *Why Althusser Killed His Wife.*
17 Mitch Rose, 'Marking a Life', in *The Creative Critics: Writing as/about Practice*, ed. by Katja Hilevaara and Emily Orley (London: Routledge, 2018), 200.
18 We are very grateful to Hanien for providing us with a copy of her original text and then for her suggestions and modifications, which greatly improved my original paraphrase.
19 Alan Drengson and Bill Devall, eds, *The Ecology of Wisdom: Writings by Arne Naess* (Berkeley, CA: Counterpoint, 2008).
20 A koppie is a small hill rising up from the African veld.
21 *Gutai Splendid Playground*, <http://web.guggenheim.org/exhibitions/gutai/> [accessed 10 April 2020].
22 Jane Bennet, *Vibrant Matter: A Political Ecology of Things* (Durham, NC: Duke University Press, 2010).
23 This extended paraphrased passage was delivered by Hanien Conradie as "*The Voice of Water: Re-sounding a Silenced River,*" (a story about collaboration), presented at the *Liquidscapes* conference, Dartington, England, 2018.
24 Perhaps significantly, the word 'art' never appeared in any official document connected with this commission, apparently as a safeguard against raising concerns about 'wasting money' from the bank's shareholders.
25 For fuller accounts of this project see: Jane Bailey, Iain Biggs and Dan Buzzo, 'Deep Mapping and Rural Connectivities', in *Grey and Pleasant Land?: Older People's Connectivity in Rural Community Life*, ed. by Catherine Hagan Hennessey, Robin Means and Vanessa Burholt (Bristol: Policy Press, 2014), 159–192; and Jane Bailey and Iain Biggs, '"Either Side of Delphy Bridge": A Deep Mapping Project Evoking and Engaging the Lives of Older Adults in Rural North Cornwall', *Journal of Rural Studies*, 28, no. 4 (2012), <http://dx.doi.org/10.1016/j.jrurstud.2012.01.001>.
26 This term is adapted from Ingold and taken here to mean a pattern of grounded identities in action who are enmeshed in multiple specific temporalities. See: Tim Ingold, *The Perception of the Environment: Essays in Livelihood, Dwelling and Skill* (London: Routledge, 2000), 190.
27 Philippe Pignarre and Isabelle Stengers, *Capitalist Sorcery: Breaking the Spell*, trans., and ed. by Andrew Goffey (Basingstoke: Palgrave Macmillan, 2011), xiii–xiv. Mary Watkins makes the exact same point in another context, see: Mary Watkins, 'From Hospitality to Mutual Accompaniment: Addressing Soul Loss in the Citizen-Neighbour', in *Borders and Debordering: Topologies, Praxes, Hospitableness*, ed. by Tomaž Gruovnik, Eduardo Mendieta and Lenart Škof (London: Lexington Books, 2018).
28 Center for Art and Social Engagement, 'Centering Creativity, Impact and Community', *University of Houston*, <http://www.uh.edu/kgmca//case/> [accessed 10 April 2020].
29 USA National Endowment for the Arts, 'Creative Placemaking Guidelines and Report Launched' (21 May 2015). See: <https://www.arts.gov/news/2015/creative-placemaking-guidelines-and-report-launched> [accessed 10 April 2020].
30 Susan Sontag, *Regarding the Pain of Others* (New York: Picador/Farrar, Straus and Giroux, 2003), 126.

7   Emergent practices

Introduction

At the risk of appearing to repeat ideas set out in the previous chapter, we must begin this chapter by returning to the question of selfhood. The phrase 'emergent practices' refers less to emergent artists in the conventional sense of a solo individual working on a self-contained practice and more to the emergence of collaborative or ensemble practices. These are practices that entangle art with skills related to various socio-ecological engagements with place and that are distinct in not particularly privileging the role of the artist over other roles. Such practices continue to question the lingering modernist presupposition that 'good' art expresses radical individuation by breaking accepted conventions. Instead, these practices re-articulate various possibilities, including those once dismissed as therapeutic, as 'redemptive play'.[1]

Socially engaged creative practitioners or local cosmopolitans often employ ensemble practices that resist the art world's globalising and hierarchical tendencies by committing themselves to working in and with specific places or communities. Turning away from exclusive 'high' art discourse and practice that continues the 'often brutal' assumption of art criticism, and that presumes 'the primacy of external judges' at the expense of the concerns of makers themselves.[2] By comparison, ensemble practices embrace a *dialogical* view of language, mind, meaning and selfhood' so as to focus on 'events occurring out in the [immediate] world between people'.[3]

These ensemble practices reflect an understanding of self as constituted in and through its relationships, attachments and connections with the world, its ecologies and eco-systems; thus questioning the still dominant assumption that artistic identity is isolated from all other activities that define 'a person's connectedness and ontological status'.[4] They challenge what Jon Thompson, the former Head of Art at Goldsmith's, identified as the 'necessarily self-centred and self-seeking' nature of the artist, qualities he considers as necessary to joining the art world seen as a 'greatly desired professional coterie and lifestyle'.[5] It is a professional coterie that, in turn, usually assumes that creativity or originality is exclusive to, and owned by, a unique individual, thus reflecting the larger assumptions about 'personhood, about nature and about society', that underpins possessive individualism and generates a toxically exclusive politics, economics and social organisation.[6]

The Czech poet and scientist Miroslav Holub offers an alternative, more inclusive, understanding of how artistic creativity is located. He points out that to identify exclusively as an artist or scientist ignores our multiple and entangled everyday intra-relations. Instead, he sees our engagement with art's 'subtle, at times permeating, but most of the time *confined*, domain' as always bound up with a plethora of other concerns – with our environment, society and the constellation of persona that constitute a self.[7] While the economics, ego-psychology and specialist discourse that validate 'transgressive' or 'cutting-edge' art continue to marginalise, belittle or ignore market-driven collaborative practices, they are increasingly central to attempts to take greater responsibility for the world we inhabit. The next section offers a framework within which to consider such practices, while the following sections set out examples of emergent practices.

The ecosophical self and ensemble practices

We can distinguish between the two poles of a spectrum: at one extreme are individuals who wholly identify with, or have imposed upon them, a *life-as –* as a Wife, an Academic, a Dutiful Son, an Artist – as any identity to which all other roles and concerns, if acknowledged at all, are subordinated.[8] At the other end is the self as an ensemble of multiple personae; a being-as-becoming irreducible to a single role or fixed identity. (This echoes Edward S. Casey's differentiation between a fixed position and a place understood as 'an essay in experimental living within a changing culture'.)[9] Both positions are notional, to the extent that in actuality most people sit somewhere on the spectrum in between these poles. The artistic practices considered in this chapter are each various forms of multiplicity in action, whether enacted in concert with others or as practices initiated by a single individual who typifies an 'ensemble identity'. As in so many other places in this book, the commitment to a multiplicity of perspectives is preserved and promoted.

What enables an 'ensemble identity' is an existential attention, listening or noticing (*notitia*) of particularity and difference as it appears, enabling us to see through given concepts and presuppositions into spaces between "category and experience, representation and reality, language and life".[10] This generates a transformative 'conversation' with the world, attempting 'to recover the neglected and . . . deeper roots of what we call thinking', and enabling us to move between and across competing authoritarian monologues by recognising that our thinking has 'contingent roots in particular persons, places and times'.[11] At the same time, these 'deeper roots' are reminiscent of what Jung referred to as collective interconnectedness to nature. He wrote: 'No man lives within his own psychic sphere like a snail in its shell, separated from everybody else, but is connected with his fellow-men by his unconscious humanity'.[12]

Something of this understanding is reflected in the claim that the real value of arts and humanities research does not lie in conventional research outputs. Rather it lies in being 'carried by and in persons' as "expertise, as confidence, as understanding and orientation to issues, problems, concerns and opportunities, as tools and abilities", and is best seen as residing in 'the notion of responsiveness'.[13] It is this aspect of 'response-ability' that thrives in conversational spaces that generate 'opportunities for discussion, argument, critique, reflection' that support modes of being-as-becoming, so that *'collaboration'* becomes a basis *'for evaluation'*.[14] Such evaluations are embodied and take place in specific locales. The geographer Doreen Massey suggests the space involved may be understood as 'a simultaneity of stories-so-far'.[15] Alternatively, these stories can be thought of as conversations between on-going evaluative narratives. This is a polyvocal process related to the dynamic materialisation of continuing collaboration and evaluation (human and otherwise), an entanglement of intra-related beings and places.

We would stress that what you have just read should not be taken as in some way dismissing those who identify their work as that of one artist. What matters is not membership (or not) of a professional coterie or participation in a particular lifestyle, but the capacity for *notitia*, the inclusivity and the complex understanding of the place of art in the wider world that a person's work manifests.

Two ensemble practices: Luci Gorell Barnes and Cathy Fitzgerald

In Chapter 3, we referred briefly to Luci Gorell Barnes' *Atlas of Human Kindness*, one element of an ensemble practice that interrelates skills, interests and an ethos at the intersections of participatory and personal arts practice, social activism, research and community engagement with specific sites (notably, in Luci Gorell Barnes' case, through the 'Companion Planting project' (Figure 7.1) that she runs on an allotment linked to a local school).[16] Central to this practice is a concern with learning and commitment to people who find themselves on the margins, particularly young children with learning difficulties, migrants and refugees. Its overarching aim is to develop flexible and responsive processes that allow us to think imaginatively with each other. She earns part of her living working in education as artist-in-residence at Speedwell Nursery School and Children's Centre in Bristol, England. This role provides one nodal point of her practice, grounding her work in, but not restricting it to, a specific geographical area and community. She can be described as a 'local cosmopolitan' (our term for those whom Bruno Latour sees as engaged in a genuinely Terrestrial politics), someone able to adopt a critical solicitude towards her locality through possessing a 'global sense of place' whose inclusive practice is

Figure 7.1 Harvesting Tomatoes, the 'Companion Planting project', Bristol. Photo: Luci Gorell Barnes.

animated by self-awareness as an informed and plural subject playing many roles, rather than being parochially framed by identification within a fixed position.[17]

Cathy Fitzgerald trained and originally worked as a biologist in New Zealand before taking art degrees in Ireland. In 2008, she began to engage with the Sitka spruce plantation she re-named Hollywood, initiating an open-ended, relational art/forest project within and with the planation that involves foresters, the local community and the constituencies of environmental politics. She combines forestry, art filmmaking, blogging, green political activism and writing in what she refers to as an eco-art social practice. She lives in the small plantation that she is working to transform, in County Carlow, Ireland, and all her larger concerns reflect her long-term commitment to this one place.

Ireland has the lowest proportion of deciduous trees in Europe after Iceland and Malta. While it has extensive but fragmented and disjointed forestry policy, addressing everything from water quality and archaeology through to biodiversity and the conservation of the freshwater pearl mussel, it shows little understanding of complex underlying issues such as the relationship between appropriate tree cover and the urgent issue of pluvial flood management. The immediate context for Cathy's transforming the plantation into a sustainably managed mixed species wood is the tension between piecemeal official policy and grass roots public interest in planting sustainable forest that includes broadleaf native tree species. However, while the sustainable management of the forest is her focus (work regularly assessed

Figure 7.2 Cathy Fitzgerald, Artists from the Wexford Cowhouse Studios visiting Hollywood in Autumn 2017. Photo: Cathy Fitzgerald.

by the Irish Council for Forest Research and Development), Cathy Fitzgerald is also engaged in a mesh of projects that set out to build educational links between silvicultural specialists, local communities, timber users, artists and environmental enthusiasts (Figure 7.2) to further eco-cultural, scientific, economic and green policy concerns locally, across Ireland and internationally.

The orientation of Cathy's activity is simultaneously ecological, creative, political and educational. It is cross-referenced through extensive personal interaction and strategic use of social media aimed at multiple constituencies. Her intention in cross-fertilising forestry with creative film work, writing and political action is to encourage exchange between diverse constituencies to provoke ecosophical thinking. Her own public self-education as a forester creatively sets out to mesh together innovative forestry practice, new conceptions of the nature/culture relationship and fundamental issues of community and environment – thus offering new ideas and models to a variety of lay and specialist constituencies.

Monoculture Sitka spruce plantations are ecologically toxic and aesthetically unattractive. An animating context here is a negotiated aesthetic that balances the economics that originally created the plantation with local desire to re-establish broadleaf native trees. However, Cathy Fitzgerald's work is also catalysing multiple exchanges between the wood's transformation and wildlife, local people, silvicultural specialists, timber users, artists and environmental enthusiasts. These exchanges are simultaneously ecological, creative, political and educational. The resulting mutual and public co-transformation is re-positioning forestry practice, notions of community and of environment for a variety of constituencies, both human and non-human. It may also assist in re-positioning notions of art within ensemble practice.

As these two examples suggest, the specific tensions and cognitive dissonances that inform ensemble practices are as different as the places from which those practices are finally inseparable. While in a literal sense Cathy Fitzgerald 'married into' County Carlow (her husband Martin took her to Hollywood on their first country drive and jokes that it was the trees she fell for, rather than him), she adds that her work there is, in part, inspired by the forests she misses in New Zealand. Similarly, Luci Gorell Barnes has had to grow into her place. This process is movingly attested to by her illustrated text *This long river*, a thoughtful homage to her elderly neighbour Jean that tells how listening to Jean's stories enabled Luci to feel that she now belongs in the valley where she lives.[18] Issues of belonging and home bring us to our next artist.

Leena Nammari

Nostos is the Greek word for returning home after a long and arduous journey. It is applied to Odysseus, for example, in the tale of his long journey home as told in *The Illiad*. The epic journey of a legendary hero is hard won through personal trials, tasks and suffering. The word has other connotations as well, such as that found in the same root in the word *nostalgia*, meaning the longing for something in the past, the wistful and bittersweet remembrance of a time gone by. Put together, these associations link the longing, the simultaneously sweet and painful remembrance of a place in the past to which one cannot return by virtue of changes and time intervening.

Nostalgia has another component as well. In Greek terminology, *algos* is 'suffering'. In combination then, the yearning in *nostos* and the suffering in *algos* create a doubly intensive pain that is afflicted when one pines for the return to home that one has left behind.

There can be no question but that this yearning is a particularly poignant malady of our time. With displaced peoples around the world, there has been no greater movement in human history than at the current time. War, famine, drought, fear and repressive regimes have wrought a diaspora of displaced humanity struggling to find a place, to remake *their* place, in the world.

Artist Leena Nammari is one of these people. Palestinian by ancestry, she and her family and compatriots mourn a loss of their homeland, their villages and their houses (Figure 7.3). Her work addresses this eviction from Ramallah, and the seamless fabric of memories from her childhood where the sights of stone buildings, the smell of the plants and the heat of the sun were as familiar as the lines on the palm of her hand.

The awareness of memories receding in the past holds her firmly now, and although she is fully aware that the desire to capture these memories is trying

Figure 7.3 Leena Nammari, *It Will Live* and *Haneen*, an installation, cyanotype
   prints on rice paper, 2018. Photo: the artist.

to hold on to an image which is ephemeral and fading, it is at the same time
imperative to her that she uses her artwork to represent the past, to see back
through time, as it were, the effects of the removal of people from their homelands, and by extrapolation, to encourage the viewers of her work to value
the 'little things' that make a home a home. The trees, the light and shadows
that falls just so on the street, the perfect arch of windows, or the sound of
flapping laundry drying on the line convey a gripping sense of place.

This evocation is, of course, to 'a placeless place', the home to which one
cannot return, precisely because (in part) it is in the past, in the child's eyes,
lit with a golden light. It is a Proustian dilemma – and for her and countless
other Palestinians, it is a home that other people now occupy. The placeless
place is nonetheless the locus of desire, and all immigrants know to some
degree that there is no return to one's own past. But that knowledge does not
obviate the longing.[19]

In academic circles, nostalgia is mostly considered unworthy of critical
attention. It frightens academics who feel more comfortable on rational
ground and less certain when emotive topics are woven into consideration.
How courageous, then, to focus upon the shunned but pervasive common-
ality in the effects of displacement; how brave to avoid sentimentality and
seek the hard and tangible material means to represent loss and the passing
of time, as an artist, and as a place-based researcher.

With each of her prints, her beloved cyanotypes that have an appropriately ethereal quality, in the crumbling unfired red ochre clay keys that each represents a 'lost' village in her installation *Haneen*, a word in Arabic that means the same as *hiraeth* in Welsh, as *morrinas* in Galician, and *saudades* in Portuguese; the pervasiveness of this longing across the world attests to its shared effect.[20] Nammari evokes the advent of 'placeless places' firmly lodged in our collective memories and in our hearts.

Siri Linn Brandsøy

Environmental change brought about by uncontrolled industrial expansion is finally encouraging empathetic re-evaluations of ways of life that were regarded, until very recently, as backward and anachronistic. Many of these ways of life were, however, much closer to being environmentally sustainable than the industrial mass culture that has effectively destroyed them. Significant contributions to this re-evaluation are being made as a result of embodied creative dialogues between the arts and anthropology.

Siri Linn Brandsøy, who has a Master's degree in social anthropology, works as a multi-media storyteller, writer and researcher involved in deep mapping, and is one half of the Sunken Bank collective. One of her long-term concerns is the creative archiving of her ongoing relationship with the island of Indrevær, one of four islands sited off the western Norwegian coastline, and the childhood home of her maternal grandfather (Figure 7.4). Continuing the *Lines between Islands 1* project that resulted in her short film

Figure 7.4 Siri Linn Brandsøy family photograph, ca. 1956/57 from an album in her grandfather's childhood home used in *Lines between Islands 2*. Photographer unknown.

Barndomshjemmet (Childhood Home), she is undertaking an in-depth exploration of what was once a significant hub for Norwegian fishermen. Home to a fishing community of some 70 people until it was undermined by urbanisation and industrial fishing, the island is now occupied, with a single exception, only for a short period each summer, by visitors who either, like Siri herself, return briefly to an ancestral place, or who holiday there in second homes.

Siri's great grandparents are buried on Indrevær, where she learned boat-lore, how to catch crabs and developed a love of the ocean. For her the island constitutes, and is constituted by, a mesh of different traces that allow her to work with the productive tension between personal remembrances of another way of living and a social anthropologist's understanding of the transformation over time of a small, close-knit community. Central to her memory-work is her relationship to the one remaining *heilårsbuar* (all-year round resident) on Indrevær, the 86-year-old Einar, and to the storyteller Jon from the neighbouring island of Nautøy. Through these relationships she has been able to engage in a variety of acts of recovery, for example with regard to specific local place names that help reanimate a world unrecorded in the official cartographic record. While Siri's situation obviously differs in many respects from that of Leena Nammari, it also has an important element in common; the loss of a once-vibrant community as the result of intervention by powerful socio-political forces external to that community. In both cases, what is being undertaken are creative attempts to keep alive "local memory and imagination as a reservoir of meanings, truths, and possibilities for a different future", acts of creative resistance that imaginatively maintain and even repopulate place in the face of "the liquidation of particular memory and local imagination".[21] For both Leena and Siri's work as well, there is a parallel to the work of Susan Hiller, whose career as an artist trained in anthropology has often addressed the capture, reconstruction and permanent loss of things that test the ability of representation, of languages no longer spoken, for example, or the words of the last speakers whose spoken thoughts fall on ears unable to hear or understand.[22]

Consequently, projects such as Siri's offer points of resistance to authorities located outside communities, and with little or no grasp of their traditions or lived experience, who deploy terms like 'resilience', 'social capital', 'community identity', 'place attachment', 'community cohesion' and 'community participation' so as to manage communities' futures. These are authorities whose values are derived from that section of society who are most expert in enhancing their own resilience to the negative socio-environmental consequences of the very system from which their social standing, authority and wealth spring.

Gareth M. Jones

Straddling the globe, Welsh-born artist Gareth Morris Jones, is a man who needs to be defined more as a verb than a noun. The term 'peripatetic' scarcely does him – or his practice – justice. He lives in Osaka, is completing his PhD with the University of Dundee, Scotland (where he did his undergraduate studies many years ago), earned an MFA from University of Tennessee and a Masters in Linguistics from Birmingham University in England.

Gareth has a walking practice through which he not only traverses ground but with which he embeds a kind of cross-cultural, chance-rich set of entanglements. Through walking, Gareth encounters the city, the landscape, people, events and objects. In doing so, he explores the self as actant, as observer and as interpreter. It is the boundary between these categories, more porous than usually recognised, and their permeability, that fascinates him.

Mythogeography is one term that is applied to this kind of investigation. Jones has been influenced by the methodology-cum-persona of Phil Smith, but his own Welsh, American, Scottish, Japanese, Malaysian and Chinese experiences and his alignment to theorists like Jane Bennett and the New Materialists, inform his own mythogeographic techniques and methods.[23] He also incorporates several aspects of Asian, and specifically Japanese, poetry and even traditional garden design with its notion of 'borrowed scenery', *shakkei* in Japanese and *jie jing* in Chinese as detailed in the *Yuanye* ('The Garden Treatise').

'Borrowed Scenery' has for a long time been understood as a principle of garden design whereby from a certain fixed vantage point an element of landscape outside of the garden, such as a mountain or a distant pagoda, was framed by a component within. Recent re-readings of the original *Yuanye* by scholars such as Stanislaus Fung have suggested that borrowed scenery advocates a readerly drift, a *shuttling between* the foreground and background, that entangles text and place, the act of reading and the act of walking.[24] 'Borrowed Scenery' moves from being a principle of design to an emergent encounter with Chinese cultural memory. This entanglement of text, place, reading and walking suggests possibilities for critical engagement with contemporary urban settings and Jones' ongoing project in his annual *Widdershins Osaka* walk.

Jones' annual group walk project, *Widdershins Osaka*, has grown out of his interest in the *Terminalia* event in Leeds in which a group walks the six boundary markers of what was once the medieval town.[25] The stones that mark this boundary are still visible within the fabric of contemporary Leeds. As well as celebrating the city's history the event also confronts issues at the heart of contemporary urban experience, such as the private ownership of public space and the imposition of arguably undemocratic civil ordinances within the city. In 2014 when he first discovered the walk in Leeds, Jones was unable to participate due to commitments in Japan. Disappointed, he then decided instead to invoke Situationist methods and bring the Leeds walk to Osaka. To achieve this, he mapped the Leeds boundary stones onto central Osaka and walked the circuit with friends. Together, they attempted to 'find' or, perhaps more accurately to discover through invocation, the absent markers through games of chance and discussion simply by looking

(a)   (b)

Figure 7.5 (a) and (b) Participants on Widdershins Dundee (2018) (a) hunt for tigers:
(b) Invoking a *'Dundee Becoming'* by recreating an Edwin Wurm-inspired
One Minute Sculpture). Both photos: Gareth Jones.

Figure 7.6 Gareth Jones introduces *ZoOsaka+: Widdershins Osaka* (2018) at the
start of the walk (centre), 2018. Photo: Chikako Ueda.

at the city around them. Their attempt to find the Leeds markers in Osaka
was absurd, of course, but succeeded in disrupting their habituated ways
of encountering space and enabled them to see the city anew. What was intended as a one-off intervention has now become an annual event: four participants in 2016 and 13 in 2018 joined Jones in this ficto-mythogeographic
drift through Osaka's *terrain vague* (Figures 7.5 and 7.6).

Ciara Healey and Adam Stead

In the context of ensemble practices, it may no longer make sense to claim that *Hold, Test, Empty, Remove, Repeat* (hereafter HTERR) (Figure 7.7), a 21 minute film produced by Ciara Healy in collaboration with visual artist Adam Stead, is a work of art, while *Already the world: a post-humanist dialogue*, a two-person performed presentation, educationally framed, that draws on an academic practice is not.[26] Both works are products of a two-year correspondence that explores the relationship of care to agriculture and place. Adam's letters address the socio-political and socio-ecological impacts of increased industrialisation and consumerism on agriculture within rural communities, framed by his relationship with the farm he grew up on and its future. Ciara's, by contrast, draw attention to the mythologies, histories and ecologies of place and their impact on our sense of belonging. Both evoke a verbal/visual dialogue and share a good deal of common material as they attend to diverse ecological, agricultural and environmental ways of knowing.

Ciara herself, who has subsequently left her university job in the UK to move back to her native Ireland, suggests one answer to the question about the status of their collaborative work:

> In many ways the work I have been doing on our cottage is an extension of my practice. . . it is a pragmatic and conscious attempt to step from the wreckage of industrial edu-business, individualism and capitalist values to make or capture interventions *into* the world rather than simply revealing its destruction. The more interventions I am involved in, through education, curation and now as a novice builder, the less I feel that sense of inertia and hopelessness that change might be possible.[27]

Figure 7.7 Healy and Stead, Still from *Hold, Test, Empty, Remove, Repeat*, video, 2018. Photo: Adam Stead.

It would be easy, but ultimately inaccurate, to identify this as the 'place-making' proposed by contemporary curators, with 'aggregating and en-abling' the ambitions of 'leading artists and arts organisations' so as to make "a difference to audience via experience and participation" and to the "arts ecology in a meaningful and sustained way".[28] As her statement suggests, she and Adam's work questions art world ambitions that require something of the conceptual distancing from its audience that characterises the ethnographer as 'participant observer'. Rather, they act as 'observant participants' in their respective places, bringing their life experience into the cultural realm through a shared ensemble practice that draws on *notitia* to inform both art and education.

Both works articulate a conversation that takes as its starting-point Arthur Koestler's notion of a machine capable of separating out the spectra of society into a set of separate, statistical units. The exchanges focus on the increasingly dislocated relationship between the suburban/urban and rural agriculture practices, particularly the mechanisation of dairy farming. They are animated by a conviction that society must reflect on rural practices and challenge the presupposition that the rural is simply stock, standing-reserve for urban/suburban consumption.

In the context of one of this chapter's concerns, to encourage a broader understanding of the role of arts practices within an ensemble practice, it is significant that Ciara concludes *Already the world: a post-humanist dialogue* with the following statement, addressed to her collaborator:

> I think in many ways the fact that we are bringing our experiences of ru-ral life into the realm of culture, Adam, questions the assumption that both worlds are exclusive to one another. As 'bridge-makers,' we turn a perceived divide into what Isabelle Stengers calls 'a living contrast.' Engaging with this living contrast is how alternative communities can be made, but we both know it isn't an easy path to tread.[29]

Courtney Chetwynd

The far Canadian North, often known as the Northwest Territories, is now more commonly called the Denendeh.[30] Artist-researcher Courtney Chet-wynd, a lifelong resident of Denendeh and Nunavut regions, has a unique practice that is shaped by the land and its peoples. She bases much of her work on the oral tradition of the Indigenous communities, learning through storytelling and through wisdom translated in the tacit meanings embedded in anecdotes and first-hand experiences. She says of her work:

> I am motivated to undertake this work because of how passionate[ly] I feel about its purpose in giving voice to invisible aspects of culture and place, while also highlighting the relationship we have to our local landscapes . . . wishing to tell a larger story about a spirit of place. . .[31]

She explores connections in her situated practice by embracing a subjective framework of land-based activities, listening to the sentient land itself and creating interventions with natural materials. Her work focusses on questioning concepts of tacit knowing, orality through making and spiritually derived knowledge. This involves relationships and reciprocity, and land-based knowing that is articulated through artistic creation in an exploratory approach towards materials and environment. Drawing upon the inter-relationships between people place and things in studio experimentation, and inner ways of knowing as methods to inform her practice-led research, some of Chetwynd's art resembles the vestiges left after shamanistic events. But equally, she deploys drawings, projections, sound recordings and audience participation in installations to convey the sense of what she calls 'invisible knowledge'. By this last, she means that the artworks may serve as a kind of conduit in leading people to their own discoveries of 'intangible knowledge'.

Throughout her artworks, she creates a material and physical investigation into ways of knowing that keeps changing its pronouns: she relies upon her own first-hand experiences to arrive at an epiphanic state of knowing.[32] Moving beyond the 'I' of the first-person translations, she listens and responds to the stories of others, critically aware and subtly preserving the ethical status of the teller as speaker, mindful of the caution of the elders and the past mistrust of a peoples misunderstood by 'academic researchers'. However, knowledge is in the ear of the listener, as well as the eye of the beholder. Her listening is not a passive listening, but an active, informed creative and performed response to hearing. And finally, she moves to address the Other, the Unseen Reader or Listener, who is there perhaps in imagination as she works, or lodged as the invisible presence within a wider community.

As a mirror to these actions, she reflects on stages of her research as groups of 'practical' activities: 'harvesting', 'trapping' and 'braiding'. This layered and multi-dimensional use of words like 'practice', 'practical' 'praxis' and so on, begin to chime with the close attention of a maker of who wants to challenge us in our too-comfortable definitions, moving out of a kind of habitual complacency into a more attentive mode, asking us to pay attention to how we live in our own environments. And by valuing the oral tradition so common to Indigenous communities and ways of knowing in Denendeh, she enacts a model for an artistic method situated in harmony with her land and her community (Figures 7.8 and 7.9).

Seila Fernández Arconada

Seila works as a multidisciplinary artist and researcher exploring various intersections of artistic methods, research practices and new social approaches. (Indicatively, she is both an honorary member of research staff in the Department of Civil Engineering of the University of Bristol, UK and a

Figure 7.8 Courtney Chetwynd, from 'landsleeper', 2-channel video, sound, salt, interface fabric, (2017). Photo: Ricardo Bennett-Guzman.

Figure 7.9 Courtney Chetwynd, *landsleeper*, video still. The artist in her environment, 2017. No photo credit. Photo: Courtney Chetwynd.

research collaborator at 'Art, Research and Feminism' at the University of the Basque Country, Spain, the country of her birth). An inveterate traveller, she has studied, exhibited work, made interventions and delivered workshops in China, Columbia, Ireland, Jordan, Latvia, Mauritius, Mexico, the Netherlands, Spain, the Ukraine, UK and the USA. Within the context of this chapter, she represents a necessary counterweight to the figure of the local cosmopolitan in the cultural ecology, acting as a highly mobile catalyst and experimenter interacting with and helping to nourish a multitude of creative place-related networks.

A recent example of Seila's approach can be seen in relation to *Afluents*, a collaborative project in which the river Mèder acted as a symbolic space where the past and present of local industrial, environmental and social concerns relating to Adoberies, an abandoned neighbourhood in Vic, Catalonia, meet (Figure 7.10). Enabled by a three months residency, she collaborated with the A+ collective, the contemporary art museum ACVic and neighbourhood associations from Calla, El Remei and Horta Vermella, among others.

Focussed on the neighbourhood of Adoberies, where ruined buildings from the leather industry still evoke an industrial heritage celebrated by local people, despite the environmental damage it continues to cause. The Mèder (a tributary of the Ter river) flows through the area, collecting and carrying a range of pollutants from heavy metals and pigments from the previous leather industry to nitrates and other contaminants produced by the pig industry (in August, 2018 Spain had a population of 50 million pigs, outnumbering its human population by three-and-a-half million). Local mechanical slaughterhouses providing pork to a number of countries kill an

Figure 7.10 Converses amb el pruner. Artistic intervention on the river Mèder. Documentation of the process taken by ACVIC Archivo. 4 July, 2017, Vic (Spain).

average of 30,000 pigs per day, with significant consequences for the environment, including the contamination of all local aquifers.

Seila initiated the workshops *Adoberies: reactivation and collaborative processes* where participants could experiment with participatory and collaborative processes, discuss ideas and try out hands-on collaborative exercises in a process to settle a common ground to work from collectively.

The first stage of her residency concluded with the event *Conversations with the plum tree*, a banquet shared with the ducks and pigeons that inhabit the river. A red tablecloth, the colour of which reflected the level of nitrates analysed in its water, was placed on top of the dam in the river, linking it to the plum tree from which participants had made plum marmalade for the banquet. At one end was food for the local birds and at the other, marmalade and pancakes for humans who discussed local identity, heritage, pollution in the river and the pig industry, which generated some controversy.

During the project a number of intertwined processes took place including a participatory documentary in which a camera pack and instructions travelled back and forth across the river (considered a border by locals) allowing stories to emerge becoming part of a collective view emerging from immersive stories ranging from the intimate experiences of local people to activists' accounts of local history. This continued with a number of events linked to the production of a local visual archive named as *Social archaeology of the river Mèder*. Meanwhile, there were a number of local environmental initiatives seeking to provide tools and knowledge to local people so as to create a neighbourhood group that could support environmental initiatives and care for the river.

The project resulted in a collective intervention named *Afluents*. This collective decision of temporarily occupying a terrain next to the river (for many years a petrol station recently demolished) presented all the work done to the further audience. It was contained in a number of collectively designed structures; symbolic shelters for all the ideas and experiences aiming to overcome the border in space (the river) and time (beyond the residency).

Laura Donkers

Laura Donkers works as an artist who believes that art does not have to be separate from everyday ordinary activities. She is convinced that art can be in itself a supportive tool for taking notice of the world and for responding in particular ways that enhance aspects of everyday living.

One of her many projects that display this is a series called *Local Food for Local People*, a Scottish government-funded series of grants for her community on the Hebridean island of North Uist, off the western coastline of mainland Scotland. In these sequential grants, Laura was first interested in the effects of climate change, of changes in gardening practices over the years and of the 'food miles' engendered by island inhabitants and the carbon footprint that this entailed.[33] She involved local schoolchildren as her 'co-researchers', asking them to engage in conversations and questions whenever they encountered someone in a garden: what were they growing? What part of it was edible? When did they harvest? Was this different than in years past? What could be grown now that was never grown before and what was no longer grown? In this way the children, and soon their parents, became engaged in a kind of phenology, a citizen science of meta-observation related to growing patterns, not dissimilar to the practices cited in Chapter 4 with Christine Baeumler's phenology work. One project led to the next: through activating the community, people came together to build a collective greenhouse, then to host growing workshops and then cooking lessons with produce that may have been unfamiliar. The community works together, and expanded the items grown for local consumption, and donating some of the bumper crops to island's families who were grateful for the food donations.

Through her practice Laura believes that art can be a supportive tool that catalyses sharing societal actions, improving the lives of others. She says that 'artists who work in this way use their art as a relational activity to advance their own personal and social transformations'. This type of practice is effectively an ensemble practice, involving the challenges of social interaction and sometimes help from groups for the benefit of individuals in a manner that means that artistic actions are no longer seen as just inconsequential aspects of daily living; the role and function of art is one of social practice created by artists working in the public realm of politics, environment and social life.

Currently, Laura is investigating traditional Maori culture and belief systems, to bring this back to Scotland and especially to her own island community, where a deeply held commitment to interconnectedness binds the people to each other and to the land inextricably. In this way, as an artist, she is aware that she too is connected in the web of relations, understanding and work with her neighbours in her community by living with them and learning from them. This requires adopting a listening paradigm so that the voices of others can be heard aiding recognition and the airing of complex issues from an insider's perspective. This spawns a sense of empowerment in those who participate.

She believes that through this affiliation a new kind of self emerges that is multiple, and expressly intertwined with the Other. It is a dialogic relationship that acknowledges interconnectedness and interdependence through projects that promote the understanding of links across the whole ecosystem. Communities with long-standing relationships to their surroundings are able to create meaningful futures where sustainability of both the environment and its inhabitants are enhanced. This facilitates long-term regeneration and restoration, challenging the politics of social isolation to find the means to build more liveable futures: a 'making-with' practice, '*sympoesis*' rather than '*auto-poiesis*', or working alone. Essentially, this is the multiple-perspective put into practice, and Laura sees her own creative practice as

Figure 7.11 Laura Donkers, *Drawing Tree 3*, Working in-situ on the Kauri, New Zealand, 2018. Photo: Laura Donkers.

an inextricable element in this web of interwoven people, place and things (Figure 7.11).

Elika Vlachaki

Elika Vlachaki is a Greek national living on the island of Aegina, an hour's ferry ride from Piraeus, the large harbour of Athens, in the Saronic Sea. She is an artist, philosopher, historian, teacher and community organiser. As a sculptor, Elika is highly attuned to materials of all kinds, and to the means to work with materials. She creates intuitively, seeing possibilities before she can even articulate why those particular materials have moved her to the making process. In her critical thinking, Elika has moved from making solitary objects in past years to a more sustained and dedicated thinking about the act and processes of making in her community. Every day inhabitants pass by the largest building that stands on her island as they go about their business; as a child, her experience of this building brought her fascination and fear because it housed political and criminal prisoners. In fact, this building is the Kapodistrian Orphanage, in the function for which it was originally intended when built in 1828, and the Prison, as it became in 1880, serving as one until 1985. As a site-specific practice, this building as a focus for Elika's investigation has been both personal as well as a deeply collaborative project as an alternative type of ensemble practice.

The Kapodistrian Orphanage/Prison, says Elika, is:

> caught between our historical associations of its 'illustrious' beginnings
> as a charitable institution and educational and cultural centre, and our
> living memory of its recent past as a prison: a painful reminder of what
> we would rather forget.[34]

In her research, she suggests that this may be the main cause of 'vexations',
the feeling of mistrust and caution that she felt when the subject of its restoration and reuse arose.

Observantly, she guessed that these ambivalent community emotions
may be a silent obstacle standing between the building's past and its future.
There was only one instance that not only did not vex anybody, but on the
contrary, brought people together: specifically, this was the church festival
when the prison would open to the community and people would go to pray
and also buy hand-crafted objects made by the prisoners.

In this historical and social context, and with the belief that with the mediation of material objects and practices through first-hand exposure people can
more easily connect not only with the past but also with each other, Elika set
off, through conversations, community events, interviews, archival visits and
books, to look for the objects that with their solid presence, their undisputed
reality, could, perhaps, do exactly that; mediate in their relationships with
each other and act, in an exhibition, as anchors – as solid points of reference
in the exterior world – for approaching the story of the Aegina Prison.[35]

Her research, then, began to focus on the material objects created by the
prisoners, and preserved by their families and the people to whom they were
sold. They became a material means by which to study the community's
memories, to learn about the stories laden with human suffering, person-
alities, joys and hopes. They became emblematic for an island's community
to speak out loud, to share in remembering, grieving, celebrating a kind of
endurance and coming to heal from socially inflicted wounds. And too, as
Sennett reminds us, they were also objects made by someone's hand, material objects that embodied the touch of their makers (Figure 7.12).[36]

David Sanders and Steve McCarthy with Henrik Melius, founder of *Spiritus Mundi*

Spiritus Mundi is a non-governmental organisation (NGO) founded in Swe-
den by Henrik Melius, a native of Malmö.[37] It is an organisation that is
dedicated to bringing people throughout the world together. Through the
powerful impact of the arts, the participants on various events and pro-
grammes in many countries have benefitted from the joy of singing, playing
music, acting and dancing. Melius believes that the release from everyday
anxieties and doubts and the promotion of expressive participation, can

Figure 7.12 Elika Vlachaki (centre left) showing schoolchildren from the island of Aegina, Greece, the objects made by political and criminal prisoners and the architectural plan of the Kapodistrian Orphanage/prison (2018). Photo: Elika Vlachaki.

lead these children to a more positive sense of self-worth and esteem. They smile, they laugh, these kids are taken outside of their narrow confines and allowed to interact with other boys and girls – a rare opportunity in some cultures – and to work closely with adults who clearly value their singing and laughter.

In a 54-minute award-winning film, produced and directed by David Sanders and Steve McCarthy of Montclair State University, New Jersey, the documentary *Hayâtuna* follows the work of two years of community interaction in Egypt and Amman, Jordan with adults, volunteers and staff of *Spiritus Mundi*.[38] They engaged Jordanian children from disadvantaged backgrounds, orphanages and those with handicaps, as they were taught music and dance. *Hayâtuna* is an Arabic word that means 'our lives', and *Hayâ*, at the root, is '*Hay*': 'that which never dies', 'the part of us that lives on'. This is an appropriate title for a creative project that was much more than a culminating final performance.

Over the two years of the Swedish-Jordanian project, and visibly in the documentary film, there is a learned confidence on the part of the children. They grow in their own competencies, in their willingness to collaborate and to take chances in order to work with other people. They learn various

Figure 7.13 David Sanders and Stephen McCarthy, *Hayâtuna* Movie Poster, 2017
      Permission: David Sanders.

types of musical forms, both Western and Eastern, but importantly, they
are given the skills to write the lyrics themselves, to understand how the
music and dances are structured. Some of the children with more problematic behaviours were patiently included in the workshops and rehearsals,
and one girl who was almost dysfunctional, anti-social and disruptive at the
beginning of the project matures into a child who transforms into a cooper-
ative and able performer. The children who featured in this project were not
only socially marginalised, but came from broken backgrounds – children
who were Jordanian, Palestinian, Syrian and Iraqi. When princes of Jordan, ministers and cultural ambassadors as well as parents and community
members, teachers and extended families, attended the final performance
the bonding that the children wrought for themselves through their performances was the true binding spirit (Figure 7.13).

   A final thought here leads one to consider the nature of the word
'community', a word that is used so often throughout this book. We have
written about many artists who work with a sense of place, who collaborate and indeed activate a spirit of community action in various creative engagements. But is it always a sense of place as the physical ground
that is shared, the proximity to neighbours in a manmade and natural
environment? It is, of course, but it transcends this definition. The ensemble practices featured in this chapter move beyond the 'sitedness' of

proximity, and into more of a social construct bound by values of, well, several things: customs, traditions, and the depth of that which is embodied in the Greek word '*ethos*'. But even more is meant here, because in this example of *Hayâtuna* and *Spiritus Mundi*, and with all of the other examples and peoples in this chapter, there is something that goes beyond the shared traditions, reaching further to embrace the complexities of living in a world that has become inexorably multicultural, with indigenes and immigrants living side by side, where multi-faith neighbours co-exist with each other and agnostics, where there is simply no one *ethos* in the '*polis*' – another useful ancient Greek word meaning city, man-made urban centre, co-dwellings.

How do we define 'community' in these terms of fuller context, then, and how can we forge meaningful 'ensemble practices'? We believe that this is achieved most effectively through recognition, through observing and valuing the customs and beliefs that we have, yes, but also by embracing – by *valuing* – precisely those who are different from ourselves, convinced that we are stronger through a multiplicity of views, beliefs and skills. In evolutionary theory Darwin made it very clear that the survival of the most adaptable was the key to survival, and, as applied to our contemporaneous world, that can only mean that it is through the valuing of ensemble, of *communitas* and of multiplicity, and through willingly embracing the need for differences to address sustainability most effectively, that we will endure. It is with this hope, this conviction, that we have focussed on these emergent ensemble practices. In the echo here of *hayâ*, it is this part of ourselves that will endure.

Notes

1 Nicholas de Ville and Stephan Foster, eds, *The Artist and the Academy: Issues in Fine Art Education and the Wider Cultural Context* (Southampton: John Hansard Gallery, 1994), 16–17.
2 Darby English, 'Modernism's War on Terror', in *Outliers and American Vanguard Art*, ed. by Lynne Cooke (Chicago, IL: University of Chicago Press, 2018), 31.
3 John Shotter, 'Life Inside Dialogically Structured Mentalities: Bakhtin's and Voloshinov's Account of Our Mental Activities as Out in the World between Us', in *The Plural Self: Multiplicity in Everyday Life*, ed. by John Rowan and Mick Cooper (London, Thousand Oaks, CA, & New Delhi: SAGE Publications, 1999), 71.
4 A. David Napier, *Masks, Transformation, and Paradox* (Berkeley: University of California Press, 1992), 21.
5 Jon Thompson, 'Campus Camp', in *Artists in the 1990s: Their Education and Values*, ed. by Paul Hetherington (London: Wimbledon School of Art in Association with Tate Gallery, 1994), 46.
6 James Leach, 'Creativity, Subjectivity, and the Dynamics of Possessive Individualism', in *Creativity and Cultural Improvisation*, ed. by Elizabeth Hallam and Tim Ingold (Oxford: Berg, 2007), 100.
7 Miroslav Holub, *The Dimensions of the Present and Other Essays*, ed. by David Young (London: Faber & Faber, 1990), 145. The italics are these author's.

8 Paul Heelas and Linda Woodhead, *The Spiritual Revolution: Why Religion Is Giving Way to Spirituality* (Oxford: Blackwell, 2005).
9 Edward S. Casey, *Getting Back into Place: Toward a Renewed Understanding of the Place-World* (Bloomington: Indiana University Press, 1993), 31.
10 Geraldine Finn, *Why Althusser Killed His Wife: Essays on Discourse and Violence* (Atlantic Highlands, NJ: Humanities Press, 1996), 172.
11 Monika Szewcyk, 'The Art of Conversation Part I', *e-flux*, 3 (2009), <http://www.e-flux.com/journal/art-of-conversation-part-I> [accessed 10 April 2020]; Gemma Corradi Fiumara, *The Other Side of Language: A Philosophy of Listening*, trans. by Charles Lambert (London: Routledge, 1990), 13; Finn, *Althusser*, 137.
12 Carl Jung, *The Collected Writings of C. G. Jung*, 20 vols., 2nd edn. (Princeton, NJ: Princeton University Press, 1970), x, *Civilization in Transition*, PAR 408. As quoted in *The Earth Has a Soul: C.G. Jung on Nature, Technology and Modern Life*, ed. by Meredith Sabini (Berkeley, CA: North Atlantic Books, 2016).
13 James Leach and Lee Watson, 'Enabling Innovation: Creative Investments in Arts and Humanities Research' (2010), <http://www.jamesleach.net/articles.html> [accessed 10 April 2020].
14 Leach and Watson, 'Enabling Innovation', 3.
15 Doreen Massey, *For Space* (London: SAGE, 2005), 9.
16 See http://www.lucigorellbarnes.co.uk.
17 Doreen Massey, 'A Global Sense of Place', in *Exploring Human Geography: A Reader*, ed. by Stephen Daniels and Roger Lee (London: Arnold, 1996), 239.
18 Luci Gorell Barnes, 'This Long River', in *Water, Creativity and Meaning: Multidisciplinary Understandings of Human–Water Relationships*, ed. by Liz Roberts and Katherine Phillips (London: Routledge, 2018), 36–53, Taylor & Francis e-book.
19 "The locus of desire" is Lucy Lippard's wonderful and evocative phrase. The full quotation is: "Place for me is the locus of desire. Places have influenced my life as much as, perhaps more than, people." *The Lure of the Local: Senses of Place in a Multicentered Society* (New York: The New Press, 1997), 4.
20 All of these words roughly translate as the desire and longing for home, a kind of 'home-sickness'.
21 Finn, *Althusser*, 145.
22 Susan Hiller, *The Last Silent Movie* (2007–08), an installation with Blu-ray 2-channel video in black and white, with sound. 22-minute audio-visual work, British Council Collection.
23 Phil Smith, *Mythogeography: A Guide to Walking Sideways* (Charmouth: Triarchy Press, 2010).
24 Stanislaus Fung, 'Four Key Terms in the History of Chinese Gardens', trans. by Mark Jackson (Paper given at the International Conference on Chinese Architectural History, 7–10 August, 1995).
25 See: Terminalia Festival, 'Terminalia – Festival of Psychogeography', <http://terminaliafestival.org/> [accessed 10 April 2020].
26 Dr. Ciara Healy and Adam Stead, *Hold, Test, Empty, Remove, Repeat*, 2018, video, 21 min., <https://vimeo.com/213206929> [accessed 10 April 2020]. In this work the dairy's milking equipment takes on a metaphorical significance.
27 Ciara Healy, in private correspondence with the author, 2018.
28 Dr Cara Courage, Head of Tate Exchange, at https://www.linkedin.com/in/caracourage/.
29 For her definition of place-making see *Placemaking and Community*, online video recording, YouTube (TEDxIndianapolis, 2017). <https://www.youtube.com/watch?v=Sfk1ZW9NRDY> [accessed 10 April 2020].
30 '"Denendeh" can also be loosely translated to refer to the concept of land ('the land of the people'). It is often used interchangeably to refer to the Northwest

Territories, which consists of thirty-three communities situated within five regions: Gwich'in, Sahtu, Deh Cho, Tłı̨chǫ (Monfwi) and Akaitcho, and has eleven official languages. . . ' Courtney Chetwynd, '"It Tells Us:" The Practice of Performativity and Liminality within Northern Culture' (unpublished doctoral dissertation, University of Dundee, 2018), 10, footnote 5.

31 Chetwynd, 'It Tells Us', preface.

32 Laurel Richardson, 'The Consequences of Poetic Representation: Writing the Other, Rewriting the Self', in *Investigating Subjectivity: Research on Lived Experience*, ed. by Carolyn Ellis and Michael Flaherty (Los Angeles, CA: SAGE Publications, 1992), 126.

33 A 'food mile' is the term of measurement used to gauge how many miles an item of food takes to be transported from its growing source to the end consumer, as a unit of measure for the fuel needed to transport it.

34 Elika Vlachaki, Chapter 2 in 'Objects as Portals: The Kapodistrian Orphanage and Aegina Prison' (unpublished doctoral dissertation, University of Dundee, 2019), 1–2.

35 Lambros Malafouris, *How Things Shape the Mind: A Theory of Material Engagement* (Cambridge, MA: The MIT Press, 2013).

36 Richard Sennett, *The Craftsman* (Cambridge, MA: Yale University Press, 2008; New York: Penguin, 2009).

> 'Material Culture' too often, at least in the social sciences, slights cloth, circuit boards, or baked fish as objects worthy of regard in themselves, instead treating the shaping of such physical things as mirrors of social norms, economic interests, religious convictions—the thing in itself is discounted. So we need to turn a fresh page. We can do so simply by asking—though the answers are anything but simple—what the process of making concrete things reveals to us about ourselves. . . .
>
> (7–8)

37 *Spiritus Mundi*, literally from the Latin translated as 'world spirit', was a term used by W.B. Yeats in his book *A Vision* (version 'A' 1925, version 'B' 1937) to describe the collective soul of the universe containing the memories of all time. In the sense used here, it may also be translated as Spirit of the World, that which binds us in an unseen but connecting way.

38 Professor David Saunders is a musician and sound artist who teaches all aspects of sound broadcasting. Steve McCarthy is a veteran international journalist who teaches video production. Together, they mutually produce, direct and edit documentary films and television features, both domestically and abroad, with their students from Montclair State University in Montclair, New Jersey. The *Spiritus Mundi* team and the *Hayâtuna project* had focussed their initial activities in El Minya, Upper Egypt. The 50% Christian town was at the time hit with several acts of violence; at one point they were in the middle of gunfire. This forced Melius to move the remaining activities to Amman, Jordan in the second year of the project. At the time, the US state department had advised US citizens not to travel to Egypt, and therefore they couldn't arrange for Sanders, McCarthy and the students to go over to Egypt to cover the activities there. It wasn't until the second year that the US team could go to Amman for filming. The editing took about three years and was finished in 2017. The film won an award in Sucre, Bolivia at the *International Human Rights Film Festival* of Sucre, August 12–17, 2018.

8   On the curation of landscapes and volcanoes

Gini Lee is many things: an academic, a teacher and researcher of landscape architecture, a designer, a visual thinker, a socio-ecological critic, a peripatetic philosopher, an environmental practitioner and a kind of farmer. In short, she is exactly the kind of hybrid creature and collaborator who is so variously featured throughout this book. Activity is one of her primary characteristics: moving, walking, travelling, visiting, investigating and applying her curiosity to far-flung places around the world and in her own home garden, all with the intention to notice. She is taken with the meeting point between the constructed and the natural environment. Many predecessors, historic, contemporary and futurist (she's a science fiction fan) can be held up as her influences: Thoreau, John Muir, Edward Casey, Rebecca Solnit, Daniel Spoerri, Emily Kame Kngwarreye, Richard Long, Val Plumwood, David Abram, Robert MacFarlane, Bill Gammage and countless others. As an Australian national, her sensibility has been shaped by the scale of the outback, by a sense of pioneer hard work and curiosity balanced by an acute acknowledgment of Aboriginal understandings, and by the knowledge that an individual's actions always matter. It is with this conviction that she applies herself to her environment, wherever in the world she by happens to be, but particularly in her home station in Oratunga, in the Flinders Ranges of South Australia. In this chapter on curation, and its application to the arid landscapes, the salt lakes, the folded ancient geologies and the dormant volcanoes of her familiar lands, the praxis of sited processes in moving and doing are paramount.

Landscape curation and the poetics of collecting – while travelling

Lee believes that:

> To curate a landscape is to work with the qualities of lands and the 'scapes that define them so to produce something novel to expand human understanding of the complex dynamics that form our experience of places – without necessarily affecting physical change.[1]

In the scope of her research, landscape, whether singular or plural, exists equally as both physical and cultural phenomena, with the ability to evoke poetic associations. Invariably, however much these places exhibit a 'natural' or 'untouched' appearance, they are somehow affected by human intervention. And in the sense applied here, although the etymology of the term 'curation' involves considerations of caring for, or caretaking of, objects and their relationships with past events and situations, such practices are usually framed by the organisational and intellectual concerns of museums, galleries and other similarly interior activities. Landscape places are mostly considered curated when they are associated in some way with museological documentation, or perhaps in making practices such as with gardens or the reconstruction of historic planned landscapes, or in sculpture parks and guided walks through human-made landscapes.

Theories of contemporary intersectionality echo the multi- and trans-disciplinary practices that are the core of this book.[2] Distinctions between designers, geologists, ecologists, performance artists, so-called 'wild' or 'nature' writers and land activists are not all clear, and may be said to be permeable divisions at most, dissolving in the necessity to address issues that cross over and connect these foci.

Mulligan and Hill trace actions in what they term, 'ecological pioneers':

> . . . Australians have made contributions like this precisely because the early experience of European Settlement in Australia began so badly. Those who were 'unsettled' by this experience often found themselves pushed into the margins of mainstream thought and practice. It has been from this position that many have been able to generate ideas and practices that have not only challenged the *status quo*, but also offered viable alternatives.[3]

Lee may well exemplify this off-centre position. She has in her Oratunga Station home several instances of these alternative practices. Her 'bush garden' is a curation in point; in collaboration with her friend, Enice, who is an Adnyamathanha ('Hills People') Elder, in the identification, arrangement, growing and placement of drylands plants for food and delight in the old Lucerne yard originally made for sheep sustenance. This small garden is an extension of her travels with Enice, by car and foot through the Flinders Ranges hills and plains landscapes, from rocky outcrops to mirage drawn salt lakes. These are the places from where various useful plant specimens are collected for experimentation, if only there will be enough water from the bore to sustain growth, as there is precious little rain. When asked if water played a role in her curation, Lee answered "that it frames it all! It comes from living in very dry places".[4] This is not just a laconic response: she continues to say that "water not only sustains all life, but that it is cultural and spiritual, that water is the magnet for everything. It is the centre of things". Given its supreme importance, she cites finding water as the ultimately critical act, along with planning around the uses or absences of water, while understanding that water is not simply a resource, since its presence brings relief and a place to pause while travelling.

At this point in the interview, she digressed momentarily, drawing a parallel between the search for 'hidden' water in the deserts and the deep drilling of bore holes in the hopeful quest of finding sweet water. This universal search has led to extreme practices elsewhere: first in Mexico by American engineers in 1958[5] as the deepest hole in the earth's surface, and then the incredible drilling project that created the deepest extant hole in the world, located in the Soviet Union, to an incredible depth of 7.5 miles beneath the surface.[6]

As an editorial aside, this desire to seek, to know, to delve into the mysteries of the earth's deepest areas Lee believes is aligned to the same human desire to find and revere holy wells, such as those found in Ireland and Scotland. During a visit with the author, Gini enjoyed a journey in the lush climate of Scotland, seeking the origins of the lore of water – trying to understand how water appearing from underground aquifers, sometimes just as cracks in the surface are fashioned into shrines of wonder. These appearances of water that spring from the ground are the discovery of the source and sustenance of life, and hark back to something profound in human evolution. Re-enacting the search and discovery is the authentication of the meaning of the symbolic sacred. Earlier in this book, we also recounted a similar shared perspective on the sacred aspect of water, citing artist Mona Smith's *B'dote* Memory Map project, '*B'dote* is a Dakota word that generally means 'where two waters come together'.[7] The *b'dote* . . . is central to Dakota 'spirituality and history', Smith reminds us.

At an instinctive level, Gini Lee appreciates this shared human impulse, but in the arid landscape where she sometimes lives, she balances this with environmental observation shared with the locals. Itinerant Italian builders made her house for the former settlers who established the Oratunga sheep station over decades, bringing with them a sensibility beyond mere pragmatic siting. Nestled between two dry creeks, with a pine-covered hill at its back, she explains the added feature, which may express a type of cosmology.[8] The four entrances to the house were constructed to correspond with the four ordinal points. Gardening on this property makes her mindful of the exact magnetic orientation, pulling to each of the four winds by unseen forces. 'The house is a beginning point for *collecting* in each of these directions', Lee explains. The word 'collecting' is stressed, leading the listener to an imagery stroll in each direction, looking for (Figures 8.1 and 8.2)

> what? a stone that feels 'special'?; a curiously shaped branch?: things less than obvious perhaps, but things and spaces to bring back to making another garden, firstly in the abstract to inform the real. An installation in a distant city seeks to ground this thinking. The catalogue notes introduce the Oratunga Project as a 'practiced place':

'A garden for untold ecologies' writes a discursive manual for making a garden for outward thinking, seeking out the here and now towards bringing ecologies back inside, through travelling leading to practicing in place. The project combines fieldwork, lists, collections, ground material, maps, conversations, walks and time travel to make an expanded garden – but is hardly a guidebook at all. . .[9]

Who has not gone walking and, say, found that a particular stone has found its way into one's hands? And by what means or why has this thing claimed its attachment upon us? Have we collected these or have they collected us? Lee's curation is a meeting point for this claiming process, this intentional attachment and arrangement, however opaque the motivating force. It may indeed be the invisible value of its ordinal direction alone when first encountered that confers significance upon the thing, a quality unknowable to us who come after. And this garden, as unusual as it may be in terms of our normal references for gardens, is made around stones collected from north, south, east and west and the small detritus that comes from fossicking in the outer landscape and in the home too.

Compass orientation, the bow to each direction, is not the sole preoccupation of the former settlers. Lee and Enice share Oratunga as a place

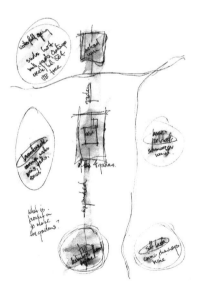

Figure 8.1 Gini Lee, *Garden for Untold Ecologies*, manual illustration, 2006. Permission: Gini Lee.

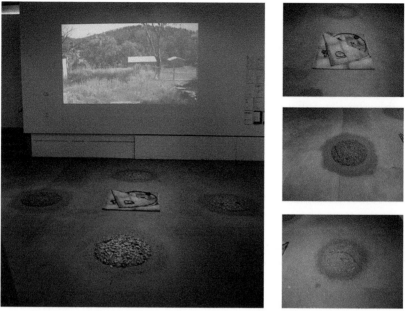

GARDENING FOR UNTOLD ECOLOGIES COLLECTED RECORDED & REMOVED TO THE OTHER SITE

Figure 8.2 Gini Lee, Installation from the *Gardening for Untold Ecologies* project, 2006. Photo: Gini Lee.

to practice, as native title, agreed over the country over the past years now allows for free access for Aboriginal people on their home country for traditional purposes. Adnyamathanha lore is shared, specifically the importance in differentiation between Northern people and Southern people. Her Aboriginal friends are Southern people represented by the 'white lichen' that appear all over on rocky outcrops in the red stone pine country. It is a designation that is a spiritual human-to-plant kinship, a kind of spiritual totem.

> A totem is a natural object, plant or animal that is inherited by members of a clan or family as their spiritual emblem. Totems define peoples' roles and responsibilities, and their relationships with each other and creation.[10]

Quite recently, her friends have suggested a relationship bound by the sharing of love for the land, by means of which Lee has been offered a 'white lichen' designation by association, in the first stage of kinship; that is, in a moiety.

> The first level of kinship is Moiety. *Moiety* is a Latin word meaning 'half'. In Moiety systems, everything, including people and the environment,

are split into two halves. Each half is a mirror of the other, and to understand the whole universe these two halves must come together.[11]

In this example as it was applied to Lee, Southern people were designated as white lichen, and Northern people were designated as 'black lichen'.[12] Notionally, this 'mirrored identity' links human and non-human, sitedness and shared life.

> Aboriginal spirituality is defined as at the core of Aboriginal being, their very identity. It gives meaning to all aspects of life including relationships with one another and the environment. All objects are living and share the same soul and spirit as Aboriginals. There is a kinship with the environment. Aboriginal spirituality can be expressed visually, musically and ceremonially.[13]

Lee expresses a cautious and deeply respectful note when speaking about the Adnyamathanha and their knowledge and care for the land who are indigenous to the region where her Oratunga home and landscape is located in the ancient, folded rocky hill country, and also of the Gunditjmara people of Southwestern Victoria in the younger volcano country of the stony rises, with whom she learned of the ancient eel farming practices described below. For these and all other Indigenous peoples, Lee has the greatest respect, and is very careful not to speak *for* them, or to misrepresent their belief systems of knowledge, or to conflate one tribal nation association with another, being very aware of their differing customs, languages and beliefs. She feels honoured by the fact that she has been 'annointed' by the Adnyamathanha, who have designated her as an honorary associate of their people. Lee has acknowledged that she has benefitted from the sharing of many stories in the oral traditions of these cultural groups, which enrich her understanding of the landscape as they are sourced from explanations of things that happened at this place or on that journey. Courtney Chetwynd in Chapter 7 echoes this same perception in her work with the Dené peoples in far northern Canada. Both Chetwynd and Lee keep the knowledge from deep listening in these stories, but they do not divulge the individual speaker or seek to claim a monopoly on the 'meaning' they interpret from the gift of their speaking. It is easier in many ways, Lee says, to relay the critical nature of fieldwork, of walking together, and locates this emergent understanding that she has gained through the active movement in the land with 'walking together'.[14] 'Sharing the country' she says, "is a process of uncovering their relations to the country as a relative". It is this language that connects Lee's, and her friends', kinship to the land, to all things on it and in it, in the act of companionate journeying. This too is the same language of totems, moiety, skin names and inter-connectedness.[15]

The practices described below offer an alternative approach to landscape curation, in works that seek to frame landscape experience through a more performative lens; one that is influenced by a community of expanded practice to create projects that are open to influences that empower the unstable, the fragile and the temporary.[16] Beyond any theoretical discourse, landscape curation projects draw from observed landscape situations to advocate enactment and/or making of new situations that attempt to reveal contemporary landscape qualities and values that do not depend upon planning leading to formal intervention.

Kate Fowle suggests that curating is a mutating form in contemporary practice[17]:

> Curators . . . take a speculative (and often meandering) approach, outlining the issues at stake from personal experience, describing a project and various artists' practices that test ways to understand key points, then making an open-ended proposition for consideration with the conclusion that research is ongoing. . .[18]

It then follows that when a mutating form of practice seeks enactment in changing spatial and material landscapes these activities have the limitless possibility for experimentation (influenced by local conditions) while simultaneously drawing from theory and precedent as digested through action. Lee's approach and inspiration for research activity leads to a curated project that finds it pertinent to employ a few useful concepts for practices affecting both tangible and intangible consequences. This listing does not infer a linear or sequential process:

- The practice of travelling
- The practice of noticing (*notitia*), leading to collection
- The practice of archiving
- The practice of inviting (ephemeral) intervention (with others when available)

To facilitate outcomes that express the possibilities of the practices above, Lee has proposed four projects for Oratunga intended to prompt interaction and intervention, whether physical, on the ground, in place, or immaterially in non-sites elsewhere, borrowing from Smithson.[19] Her tactics include but are not limited to:

1   To make a conceptual/ephemeral museum/gallery for Oratunga; one that is conceived/practiced through a series of events that are spontaneous or curated.
2   Layers of dissimilar data are reconfigured as interconnected collections. Artefacts/objects encountered and/or collected do not 'fix' space; rather they act as points or moments to move with/against.
3   Devices placed in the landscape invest in found material collections and temporal spaces facilitating the agency to make many itineraries.

4 Scale relationships dissolve where the landscape may be experienced through the curated tour, over the course of three hours, three days or any other temporal interval.[20]

The presence of water

The issue of water is present throughout all this. The presence and absence of water, of droughts and floods, of aridity and lushness depending where one travels and according to climate in Australia – known as the driest continent is *understandably*, all-important. In fact, Lee says, "water is life", in our interview, while half-focussing on several thoughts at once.[21] Certainly there can be no life without water, and although there is a dearth of water in arid country, water does exist, however sparsely. Part of the lack of visible water in the forms of rivers, lakes, ponds, waterfalls and streams is misleading, especially to those of us lucky to reside in more watery landscapes: here, in scorching temperatures, water often takes an underground path. The Great Artesian Basin, for example, which is a great pool of water underground that has been trapped by layers of sediment and geologic formations, covers an estimated 22% of the Australian continent, and contains 1,700,000 (i.e., 1.7 million) square kilometres of water. In many places, this is the only fresh water available.[22] The volume of fresh water seems to stretch one's ability to imagine: something like the equivalent of 130,000 Sydney Harbours is estimated to exist there.

There have been many surveys of the aquifer, water tables and watersheds in these districts by the Australian government. Government surveys of 1996 and 2008,[23] and the Hydrogeological Atlas project,[24] for example, have been looking for precisely these important waterways that are partially or almost fully underground adjacent to the Flinders Ranges that Lee calls home (Figure 8.3).[25]

While these are what Lee calls the 'essential part of a landscape', they are nevertheless unseen. Since records began, there has been an alarming diminution of this precious resource; an estimated 90% of what was once estimated to have been in the Great Artesian Basin may already have been lost to extraction. What remains in the figures given here, is the 10% vestige that is crucial to the continuation of life. The fact that so much of the water lies underground has been what has 'saved' this resource from recent evaporation or contamination in our current period of rising temperatures and climate change.

Water in the arid landscape has played a significant role in occupation and settlement since humans first arrived in Australia more than 60,000 years ago. Permanent sources of water were integral to maintaining the Aboriginal people within the Flinders Ranges who are generally identify today as the Adnyamathanha People. Historical records indicate six other distinct language groups had territorial connections with

Figure 8.3 Lake Kati Thanda Lake Eyre, with a little local water, 2018. Photo:
Gini Lee.

the Flinders Ranges and surroundings, namely the Pirlatapa, Yardli-
yawarra, Ngadjuri, Nukunu, Pankarla and Kunyani peoples.[26]

Lee's research into the cultural relationships between people and water
in remote arid Australia looks at the relationships between surface water,
that flows from the intermittent rains, and ground water which either is
drawn up from deep underground or bubbles or seeps up in the numerous
but sparsely located springs. There are also vast rivers, usually dry, that all
drain into Kati Thanda Lake Eyre, the lowest point of Australia, and the
closest there is to an inland sea. In this system are water bodies, ecological
refugia, that are both resource for pastoralism and tourists, and home to
many Aboriginal stories and practices. In Aboriginal lore, these places can
be gendered and, therefore, under the custodianship of women in one place,
and men in another, where they are responsible for the care of the place and
the fish that are a necessary source of food in dry times.

These 'critical refugia' are typically fresh or semi-saline waterholes located in:

> large generally dry river systems, wetlands and ephemeral lakes. The
> more permanent waters become the only refuge for isolated or relict
> populations in times of severe drought and form a network of refugia
> linked by the extensive floodplains that fringe central Australia's inland
> rivers. Their healthy condition is essential to the continuing presence of
> animal and human populations and livelihoods. If these waterbodies

become marginal, either through climatic change or due to unsustainable human use, then their ecological value becomes under threat.[27]

Ancient eel country: eel farming adjacent to volcanoes

Away from the arid centre, the more temperate Stony Rises volcanic country of southwestern Victoria reveals Aboriginal water practices in operation for millennia. The presence of seasonal standing water in the Lake Condah region has fascinated anthropologists and archaeologists for many years, especially because the Indigenous Gunditjmara people have been so closely tied to its wetlands and waterways. This area lies adjacent to the lava flows of Mount Eccles in the National Park in southwest Victoria.

> The Kerrup Gunditj clan at Lake Condah had traditionally engineered an extensive aquaculture system at Lake Condah. Other Gunditjmara clans along the Budj Bim landscape worked together to establish *kooyang* (eel) trapping and farming systems, developing smoking techniques to preserve their harvest – probably one of the first cultures in the world to do so.
>
> They continued to live and work on their country in a highly complex society until Europeans arrived in the region. As pastoralists moved further into southwest Victoria, the stone country of Lake Condah became a sanctuary for Gunditjmara people providing eels, possum and kangaroos for the families.[28]

Preserved in the stones of this area, the vestiges of the Aboriginal system for damming, stocking, fishing and trapping still remain (Figure 8.4).

Secrets of the Stones[29]

> For nearly 8,000 years, the Gunditjmara people of western Victoria farmed eels. They modified more than 100 square kilometres of the landscape, constructing artificial ponds across the grassy wetlands and digging channels to interconnect them. They exported their produce and became an important part of the local economy. And then white settlers arrived and all they left of the Gunditjmara's thriving industry were several hundred piles of stones that had formed the foundations to the people's huts.
>
> Since the 1970s, archaeologists have suspected that the stone remains in the Lake Condah region were evidence that the local Aborigines had lived in villages.
>
> The area was naturally wetland; Heather Builth discovered that the Gunditjmara had modified it with weirs, channels and dams to make the landscape eel-friendly.[30] The output from these eel farms would have been enormous – she estimates it could have fed up to 10,000 people. Her hunch was that this was more like an ancient fishing industry

Figure 8.4 The *Stony Rises* and ancient eel country, with dry stone wall by later
    settlers who farmed in other ways. Photo: Gini Lee.

than a subsistence farm, and she set out to prove it. She had noticed the
landscape was scattered with burnt, hollowed-out trees, often right next
to the eel traps.

Could the structures have been ancient smokehouses? Builth took soil
samples from the base of four trees. When the results came back it was
her eureka moment: the samples did contain traces of eel fat. Suddenly
the whole picture changed. "The Gunditjmara weren't just catching
eels," she says. "Their whole society was based around eels."[31]

The value of this work on re-establishing Gunditjmara wisdom and land
practices, so surprising in many ways, is another example of reconstructing lost knowledge that has now informed conservation policies for water
management, conservation and preservation.[32] The use of the naturally occurring stony rises, the water and a carefully managed food industry have
shaped conservation and land use policy in this region that was under the
control and advice of the Gunditjmara.[33]

Lee invited the architect and academic Steve Loo, a Chinese and Malaysian cooking aficionado, to input into her *Stony Rises Deep Mapping* project,
expanded upon below. Relating her walking to Lake Condah with Gunditjmara and other artists, together with the amazing eel farming story in this
stony place, later elicited a desire to cook eel soup, generating a tangen-
tial association across practices, lands and cultures. This was then inserted
into the deep mapping process. Curiously, the Chinese were early settlers
in these arid lands, regularly establishing market gardens on the banks of

critical refugia, a talent that often eluded European settlers. These gardens are long gone, but the stories remain.

On geopoetics and tactics for mapping and thinking[34]

Gini Lee describes her mapping works as "a stratigraphy of text – as a kind of writing over writing where points once separated in time are made adjacent – through the medium of the gridded mat".[35] Drawing inspiration from Cliff McLucas' manifesto for deep mapping, *There are ten things that I can say about these deep maps . . .*,[36] the work is focussed by a curatorial approach to guiding peripatetic travelling through a landscape and consists of arrangements of fragmentary impressions ordered and layered conceptually using "maps, writings, historical accounts, paintings, photographs, compilations, stone works, fabrics, markings, grid references, books, digital displays of sequential walks, performances and promises". *The Stony Rises Project* was developed by the RMIT Design Research Institute, Melbourne, to research and develop artistic and design works drawn from Victoria's volcano country. Gini Lee was one of the ten people selected to make a work as offering to the project, and she sought to work with both the volcanic stone country and also with the older stony country of the Flinders Ranges in South Australia to uncover the possibility for relational narratives and or aesthetics across dissimilar landscapes. Lee sees this project, begun in 2008 and ongoing, entitled *Deep Mapping for the Stony Rises*, as assembling the various topographies and topologies encountered while making a cross-landscape environment relating to six particular places within the Stony Rises of Victoria and the Flinders Ranges of South Australia. As a means to illustrate how these actions play out in projects for landscapes the following is an account of a collaboration that was unexpectedly prompted by others' curation of the *Stony Rises Project* for now-extinct volcano country (Figure 8.5).

She frames this work as on-going collaboration with others, and without completion. Two phrases here, '*without* completion' and '*for* the Stony Rises' (author's italics), need to be given their full weight and emphasis. Both are indicative of a distance between her commitment and the numerous (and sometimes facile) ways in which the term 'deep mapping' has been adopted as a catch-all by artists and others whose activity happens to involve some cartographic dimension.[37]

To recognise the seriousness of Lee's practice in this way is not simply to dismiss the appropriation or adoption of 'deep mapping' by members of a culture industry always desperate for a 'new angle' out of which to manufacture novelty. That process, which involves different degrees of comprehension, appropriation, misunderstanding, dilution, absorption and development, follows the emergence of any new praxis and then takes its own course, regardless of the concerns of any individual or group tempted to claim 'ownership' of it. Equally, the quality of Lee's research and material

(a)

(b)

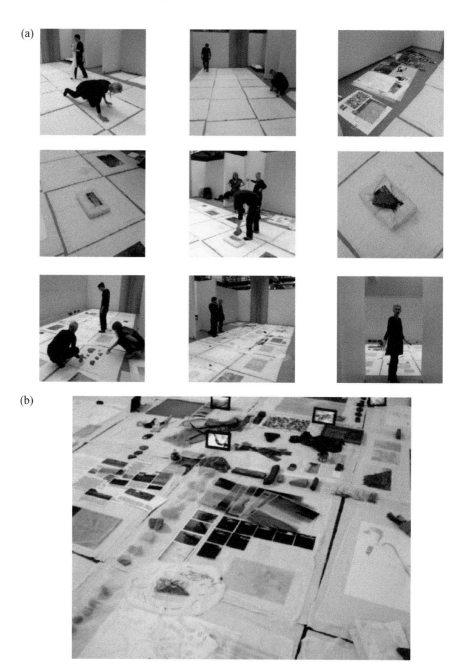

Figure 8.5 (a) and (b) Gini Lee and collaborators Linda Marie Walker and Steve
Loo: *Deep Mapping for the Stony Rises*, a contribution to the *Stony Rises*
project. Details of a floorwork assemblage, including found objects, photographic prints and digital projection, 2009. Steve Loo's embroidered
recipe for eel soup is lower left. Photographer unknown.

work is not diminished by this process of appropriation, and should not be judged in terms of any modish fad. Her research is based upon long-standing and genuine interdisciplinary investigation, years of observation and activity, and collaboration within her urban and rural communities,

In positive terms, it may be said that Lee's generosity of intellectual and creative response in her *Deep Mapping for Stony Rises* project allows her to collaborate with, and respond to, the many other peoples in the understanding of her beloved places. In fact, it is precisely through this sharing and seeing through the eyes of others, that Lee has expanded her own deep knowledge of her own rural home in South Australia, as well as the Stony Rises and the eel country of Lake Condah, and this has collectively expanded her ability to represent this deep and overlaid comprehension.

Volcano chorography[38]

Lee's Volcano work was conducted with others in a field trip of 2008, that comprised walking with the locals to cherished places of interest. The walk to the extinct volcano was undertaken by the group in concert with the local farmer, Neil Black on whose property it lies.

The Mount Noorat project (2009), was created as a scripted way to traverse the volcano and is offered towards translating the intent of Lee's proposed landscape curation.[39] Mount Noorat is an extinct volcano in the stony rises country of southwestern Victoria in southern Australia, and one day a group of artists, architects and a landscape architect were touring the rolling countryside in order to eventually make an exhibition about these landscapes. After the trip and as a response to the exhibition installation two of us (Lee and Gibson) began to converse, at distance between Sydney and Melbourne, on a way to approach such a place, for that moment and for the future. Ross Gibson devised a textual journey in remembrance of that afternoon and Gini offered a photographic assemblage specific to time and place to aid in visualising future visits should anyone stumble across the possibility of re-enacting the event. Here, following, is the textual result:

Ross: Mount Noorat
   Peering down into the inverted cone of the crater, I thought: there's a simple action that's transformative, home-made and beautiful that's available here and what I could do with it is rather like art – because of how quietly astonishing it might be.

This is how I'd do it, using just a few guiding protocols, and one basic piece of technology:

1   *Through a careful arrangement that will have been brokered already via trusted local people, secure entry permission from Neil Black. Give him ample time to make his decision.*

2   *Be completely aware of his generosity. Don't apply pressure. Don't give him any worry. Be patient awaiting a response.*

3   *Take 100 metres of heavy gauge natural rope. Lug it up there. Secure the rope to the base of a substantial tree growing at the rim of the crater.*

4   *Holding the rope, walk in a straight line backwards down into the cone of the crater, laying the rope down in front of you as you go so that it will be available to you when you will haul yourself back out of the crater at the end of your session.*

5   *Once you have reached the base the base of the crater, clear the ground (a 2 meter × 1 metre rectangle) of gibbers and tussocks*

6   *Lie flat on your back on this cleared ground and gaze up at the sky orbing above. Stay supine like that until you begin to feel the vast heavens pulling you upwards.*

7   *Enjoy the odd sense that you could hurtle out into the immeasurable light, as if powered by an irrational, anti- gravitational urge. Develop your own psychic control over this inverted vertigo. It is a contest between panic and bliss.*

8   *Enjoy, for at least 15 minutes, the sense that you are flying, not in the realm of physics so much as in the realm of your own psychic potential. Stand up and find the rope.*

9   *Haul yourself back up to the tether tree.*

10   *Leave the rope in place for everyone who comes after.*

11   *Inform Neil that you have left the rope there and offer to remove it if that's what he prefers.*

Fini.

* * *

Collaborative projects of journey/movement/walking together as a kind of 'performative curation' have an echo with other walking artists, such as the praxis of Gareth Jones in Chapter 7, as well as Hamish Fulton, Richard Long, Francis Alÿs, Janet Cardiff and George Bures Millar, Vito Acconci, Dee Heddon, Paul Smith and the Situationists, among many others. What each of these artists do in their walking practices vary from interventions, to sound-recorded journeys, to cultural encounters, and of course the whole tradition of pilgrimages, to meanderings, to walkabouts, to Chaucerian tales *en route* and stand in a long-established genre of episodic narrative and experiential confrontations. A walk, like a river journey, is so close to a temporal metaphor that it ceases to be metaphoric and becomes *emblematic* of time and space. The qualities of the land or space one traverses, that which one encounters along the way can be conveyed as an extended narrative (Chatwin's *Songlines* comes to mind, a much-contested text, or Huckleberry Finn's rafting progress down the Mississippi River), a travelogue, a coming-into-being, or a psychological journey into liminal spaces of the human psyche. This transformation from physical landscape and hardships of the journey translate easily into the observation of growing maturity and spiritual insights; the traveller is never the same as the returning walker as he was when he left. And the land traversed is equally compelling as a liminal place of *both* geographical and spiritual spaces. It is precisely this overlay of the known and the unknown, the chance encounter and the planned well-trodden route, as well as the mysterious and intuitive grasp of the unreachable in the journey that makes this trope so archetypal.[40]

The artist Robert Smithson was fascinated with this juxtaposition of dimensions and layers. In his definition of 'non-sites' he wrote:

> *The Non-Site (an indoor earthwork)** is a three-dimensional logical picture that is *abstract*, yet it *represents* an actual site in N.J. (The Pine Barrens Plains). It is by this dimensional metaphor that one site can represent another site, which does not resemble it – this *The Non-Site*. To understand this language of sites is to appreciate the metaphor between the syntactical construct and the complex of ideas, letting the former function as a three-dimensional picture, which doesn't look like a picture.[41]

Representation of the land-site in *The Non-Site* establishes a 'syntactical' connection, eschewing the anticipated mode of mimetic picturing, and instead linking 'the complex of ideas'. Thereby, it provides a means for stepping into material signifiers that can 'abstractly' connect time, space, place and journeys: an ideal trope for Lee's type of practice.

* * *

A couple of years before Lee's *Mount Noorat Project*, she wrote on recollecting Susan Sontag's words about the allure of volcanos and the nature of collecting and of the walk to the volcano, which was simultaneously dangerous and compelling,

> He was a collector of lists, of travels, of artefacts, but rarely of people: 'collecting expresses a free-floating desire that attaches and re-attaches itself – it is a succession of desires. The true collector is in the grip not of what is collected but of collecting . . .[42]

These earlier writings on the nature of collecting experience rather than of material things sound in resonance:

. . . His collector's attitude was singular, beyond mere accumulating, rarely collaborative and always competitive, although he also collected for the love of rescuing things from neglect or oblivion. Yet the professional collector simultaneously developed a fascination for the Mount Etna volcano and its eruptive qualities, and thus expanded his collector's space from that confined to the house and museum, to that of the free-floating wanderer in the landscape: although he did conduct guided tours up the slopes. He was captured by the sublime dangers in the changing landscape, and collected the responses evoked by the tours he curated, alongside physical evidence in the detritus produced by the erupting ground.[43]

Sontag's *The Volcano Lover*'s account of the landscape is highly detailed and beyond the desire to collect the stone erupted over time. He is more a curator of landscapes than a curator of displays. He curates the landscape because it is too vast, too complex and too changing to enable possession, and he collects volcano matter by both personal exploration and by following expeditions, over and over again. And each time, the familiar territory is made new again by fresh eruptions and emerging topography.

Lee evokes the performance of walking and guiding in experiencing the Mount Noorat extinct volcano together with Ross Gibson – at a singular moment in time never to be repeated – while also collecting materials for the larger Stony Rises deep mapping installation. This experimental chorography is about shared memories, new narratives and reveals an easy comfort between occasional friends in the practice of walking with unknown intent at that moment. The spirit of making a sublime narrative for this landscape enable a partly fictitious journey to be made later on; where abstraction superimposed over landscape re-constructs fragmentary moments in the ongoing life of places. The landscape becomes the locus for journeys and personal collections that transform scale, materiality and temporality through actively noticing local aesthetics in detail; the collection of experiences is inspired through uncovering traces of interaction with nature, mostly realised in the photographic record and writing but also re-imagined through storytelling.

The literature of place or 'deep travel' references a mapping form evinced through vertical movement (beyond the implied horizontal nature of cartographic mapping) which commences from the surface of the land and moves backward in time, searching for hidden social and cultural dynamics embedded in that geographic context.[44] A specific territorial reading may reveal all strata that become elements of place – self, nature, society, history, geology, ecology, politics, economy – where space for conceptual wandering and finding may be found. Other readings drawn from the practice of vertical thinking investigate spatial strata of material forms through immersion in collected narratives written, shared or imagined over time. Ideas around assemblage and the performativity of people and things inform such thinking expressed through the annotation of forms and maps drawn from personal journeys to particular geographical and geological places and territories.

Curating a landscape, as Lee has proposed, begins with the close observation that underlies an environmental practice. This has connection to what some refer to as 'Deep ecology'.[45] "In 1973, Norwegian philosopher and mountaineer Arne Naess introduced the phrase *deep ecology* to environmental literature".[46] The eight points of the Deep Ecology Platform offered in this 1995 conference at Schumacher College emphasised that all life had value; that richness and diversity were in themselves desirable qualities, entirely apart from human wishes; that the impact of humans is transgressive and worsening; that the non-human impact was excessive, especially with regard to diversity; that urgent change is needed; and that those who understand and agree have an imperative to act accordingly, peacefully and democratically'.[47]

Ecosystems in the context of families and kinship

It is worth pausing for a moment here to merge a few of the threads above. Totems, described above, are one spiritual/philosophical/ecosophical method whereby the environment and humans merge in mirrored identities. Earliest writings about ecosytems often included analogies to kinship, to families, with actions and consequences initiated by one affecting the whole.[48] Lest one forget, an important reminder here is to remain healthily 'agnostic' about the cosiness or otherwise of families and kin: for some, this is a heartwarming comparison and conjures notions of belonging. For others, however, this is problematic, and does not serve as a positive parallel and, in fact, may have precisely the opposite effect of distancing, constraining or rejecting happenstance circumstances of birth. But to set aside the analogy and go straight to the point, there is an undeniable, inextricable connection between *situated life*, being human in a place as a lived experience, and the world-as-place in which this dwelling occurs. The space of the Australian landscape in this case, or of the interior of a jug wherein the unfilled space 'is not nothing' folds function, site, physicality and reciprocal effects into one. Interweaving the elements here as water, earth, sky – the living organism upon which human and non-creatures co-exist, Lee and her collaborators closely observe and respond to the changes and qualities they perceive, drawing attention to those manmade effects that are causing such a shift in their environment.[49] For example, the drought in New South Wales (in 2018), for example, 'was the second driest on record'.[50] The gradually increasing incidence of drought, the incrementally increasing temperatures, and the spate of fires in recent years that adds to the CO_2 emissions in the atmosphere, has had the effect of focussing world attention, and urgent community actions, across Australia. If the Anthropocene bodes ill, it may at least have the effect of concentrating attention on the most minute of human actions as well as the largest of actions and their subsequent impact.

Gini Lee's research methods were described above as the highlighting the following:

- The practice of travelling
- The practice of noticing, leading to collection
- The practice of archiving
- The practice of inviting (ephemeral) intervention (with others when available)

In a process of "moving, walking, traveling, visiting, investigating and applying her curiosity to far-flung places around the world" – the phrase used at the beginning of this chapter to describe Lee's praxis, there was an implication that all of the actions were outward-looking, focussed in the external world of her beloved Australian homeland. And yet, as Elkins suggests, there is a problematic relationship with describing the environment as purely external. He suggests:

> It seems to me just possible that landscape, perhaps along with the body and its representations, is an intractable subject for scholarship, in the sense that it resists the illusion of an observing subject, situated well outside the object of study and contemplating it with the protection and support of a historically grounded series of protocols and methods. Like the body, landscape is something we inhabit without being from it: we are *in* it, and we *are* it.[51]

This is an apt reminder that we as observers are not simply neutral or external to the subject of investigation. In Lee's case, she is as happily descriptive of her own changing persona as she inhabits her station, as she is an agent of change in moving away from its sheep farming origins to a centre of land investigation located in a site as a locus for researchers, climate change investigators, government water scientists, Aboriginal visitors, international writing workshops, students, artists, geographers and poets. She curates, in the sense we have just discussed, "making an open-ended proposition for consideration with the conclusion that research is on-going". And happily so for, as a speculative human whose curiosity leads her along her path, Gini Lee is *in* the land and *of* it.

Notes

1 Gini Lee, 'The Intention to Notice: The Collection, The Tour and Ordinary Landscapes' (unpublished Doctoral dissertation, RMIT University, 2006).
2 Max Oelschlager, *The Wilderness Condition: Essays on Environment and Civilisation* (Washington, DC: Island Press, 1992).
3 Martin Mulligan and Stuart Hill, eds, *Ecological Pioneers: A Social History of Australian Ecological Thought and Action* (Cambridge: Cambridge University Press, 2001), 12.

4 Interview with Professor Gini Lee, by M. Modeen: 3 Sept. 2018 (Hereinafter, simply designated as 'Interview').
5 Project Mohole. American engineers drilled through the Pacific Ocean floor off Guadalupe, Mexico. But Congress discontinued funding in 1966 before the drillers ever reached the mantle. Alicia Ault, 'Ask Smithsonian: What's the Deepest Hole Ever Dug?', *Smithsonianmag* (2015), <https://www.smithsonianmag.com/smithsonian-institution/ask-smithsonian-whats-deepest-hole-ever-dug-180954349/> [accessed 10 April 2020].
6 The Kola Superdeep Borehole was just nine inches in diameter, but at 40,230 feet (12,262 meters) reigns as the deepest hole. It took almost 20 years to reach that 7.5-mile depth—only half the distance or less to the mantle. Among the more interesting discoveries: microscopic plankton fossils found at four miles down. The Kola hole was abandoned in 1992 when drillers encountered higher-than-expected temperatures—356 degrees Fahrenheit, not the 212 degrees that had been mapped. Ault, 'Ask Smithsonian'. <https://www.smithsonianmag.com/smithsonian-institution/ask-smithsonian-whats-deepest-hole-ever-dug-180954349/> [accessed 15 September 2018].
7 Sisseton Wahpeton Dakota storyteller and media artist Mona Smith, is creating a memory map of the *b'dote* area of the Mississippi and the Minnesota Rivers. This *b'dote* (a place where two waters come together) is central to Dakota spirituality and history. See: Bdote Memory Map, 'Bdote Memory Map: Mitakuyepi! Welcome!', <http://bdotememorymap.org/> [accessed 10 April 2020].
8 Interview.
9 Manual exhibition, UniSA Gallery, Adelaide.
10 Graham Paulson, 'Aboriginal Spirituality', *Australians Together* (n.d.), <https://www.australianstogether.org.au/discover/indigenous-culture/aboriginal-spirituality/> [accessed 10 April 2020].
11 Paulson, 'Aboriginal Spirituality'.
12 Paulson, 'Aboriginal Spirituality'. ". . . while members of one Moiety protect and conserve the animal, members of the other Moiety may eat and use the animal". Thus, Moieties are an in-built system to protect, use and conserve the animal or plant.
13 E.K. Grant, *Unseen, Unheard, Unspoken: Exploring the Relationship between Aboriginal Spirituality & Community Development* (Adelaide: University of South Australia, 2004), 8–9. E. K. Grant, a community developer from Adelaide in South Australia, identified that 'Aboriginal spirituality' is used 'by non-Aboriginals and Aboriginals alike without reference to its meaning and its roots and that 'in itself it is a self-defining entity with each person defining it within his or her own framework of knowledge and experience'.
14 There are echoes here, too, of Gareth M. Jones' walking practices in Chapter 7.
15 Paulson, 'Aboriginal Spirituality'. 'The third level of kinship is the Skin Name. Similar to a surname, a Skin Name indicates a person's bloodline. It also conveys information about how generations are linked and how they should interact. . . Unlike surnames, husbands and wives don't share the same Skin Name, and children don't share their parents' name. Rather, it is a sequential system, so Skin Names are given based on the preceding name (the mother's name in a matrilineal system or the father's name in a patrilineal system) and its level in the naming cycle' (n.p.).
16 Cliff McLucas, *There Are Ten Things I Can Say about Deep Maps– Number 10*. 'Deep Mapping', *CliffordMcLucas* (n.d.) <http://cliffordmclucas.info/deep-mapping.html> [accessed 10 April 2020].
17 Kate Fowle is the chief curator for the Garage Museum of Contemporary Art in Moscow and Director-at-Large at Independent Curators International (ICI) in New York, where she was Executive Director from 2009 to 2013.

18 Kate Fowle, 'Introduction', in *Thinking Contemporary Curating, Edition 2*, ed. by Terry E. Smith (Los Angeles: University of California, 2012), 8.

19 Robert Smithson, *Unpublished Writings* in *Robert Smithson: The Collected Writings*, ed. by Jack Flam, 2nd edn (Berkeley: University of California Press, 1996).

20 Gini Lee, 'The Intention to Notice: The Collection, The Tour and Ordinary Landscapes' (unpublished doctoral dissertation, RMIT, 2006), 225.

21 Interview.

22 It is estimated to hold nearly 65,000 million megalitres of water, (equivalent to about 130,000 Sydney Harbours) and is a key source of water for springs, many of which support unique ecosystems. See: Tim Ransley, and Andrew Feitz, 'Navigating Australia's Largest Groundwater Resource', *Australian Government: Geoscience Australia*, <https://www.ga.gov.au/news-events/features/navigating-australias-largest-groundwater-resource> [accessed 10 April 2020].

23 Published by the Government of South Australia, Knowledge and Information Division, Department of Water, Land and Biodiversity Conservation: 2008.

24 The Hydrogeological Atlas of the Great Artesian Basin draws upon previous work undertaken by Geoscience Australia that has contributed to a number of projects, such as the CSIRO-led Great Artesian Basin Water Resource Assessment (GABWRA) project and Geoscience Australia's Carbon Capture and Storage project.

25 Jones and McEntee, (1996), as cited in: Melissa White and Glen Scholz, *Prioritising Springs of Ecological Significance in the Flinders Ranges*, Government of South Australia, Knowledge and Information Division, Department of Water, Land and Biodiversity Conservation (Adelaide: Department of Water, Land and Biodiversity Conservation, 2008), 3.

26 Jones and McEntee, (1996), as cited in: White and Scholz, *Prioritising Springs of Ecological Significance*, 3.

27 Over time Lee has surveyed many waterholes in the big Cooper Creek and Neales River systems for the South Australian Government in the Critical Refugia Project, looking at relationships between culture and water in surface and ground water regimes.

28 See: <https://www.environment.gov.au/indigenous/ipa/declared/lake-condah.html> [accessed 16 November 2018].

29 Graham Phillips, <http://www.eniar.org/news/stones.html>, [accessed 31 July 2016]. This rather long quote was cited at length because the website is no longer extant.

30 Archaeologist, Dr Heather Bulith of Monash University has studied the Budj Bim (Aboriginal for 'top of head sticking out') volcanic landscape and associated swamps and lakes such as Lake Condah since before 1996. She has been mapping the Lake Condah area and documenting its archaeology. The project is part of a long-term study to analyse and interpret the cultural and environmental landscape of the lava flow.

31 Graham Phillips, 'Secrets of the Stones', *Sydney Morning Herald* (13 March 2003), <https://www.smh.com.au/national/secrets-of-the-stones-20030313-gdgf3f.html> [accessed 10 April 2020].

32 The Kerrup Gunditj clan of the Gunditjmara 'dug a series of channels, weirs and ponds in the wetlands. They also placed woven eels baskets in gaps in the weirs to catch the eels; other baskets were used to divert the eels downstream once the ponds were full. The eels were then captured, placed on sticks and smoked inside hollow trees. This year-round supply of food meant the Gunditjmara gave up their nomadic life, building stone huts alongside their eel farms and within sight of the revered and now-dormant Budj Bim volcano in western Victoria'. Paul Chai, 'On a Mission: Uncovering the Past of Victoria's Gunditjmara Country',

Traveller (2017). See: <http://www.traveller.com.au/on-a-mission-uncovering-the-past-of-victorias-gunditjmara-country-gtvi9t> [accessed 10 April 2020].

33 See: <https://www.pmc.gov.au/indigenous-affairs/environment/indigenous-land-and-sea-management-projects> [accessed 16 November 2018].

34 The reference for 'Geopoetics' here and elsewhere in this book do not correspond to, or reference, The Scottish Centre for Geopoetics, which was founded in Edinburgh on Burns Night in 1995 in response to various texts by Kenneth White. While there are obvious points of overlap and mutual concerns, the distinctive way in which we authors use this reference has more to do with the associative processes of 'poetics' as applied to the natural environment and its care, and less to do with alliance to any single organisation.

35 <http://thestonyrisesproject.com/the-exhibition/the-works/> [accessed 31 July 2016].

36 Cliff McLucas, 'Deep Mapping', *CliffordMcLucas*, <http://cliffordmclucas.info/-deep-mapping.html> [accessed 10 April 2020].

37 The need for openness does not preclude focus on particular concerns. This is clear from Lee's referencing an email from Paul Carter as presenting "the point" of her research that could be used to define a certain approach to deep mapping. Carter wrote:

> But essentially: aren't you interested in developing an amplified notion of cultural and environmental heritage? Isn't a preliminary to this the design of a system of environmental notation that gives recognition to temporal traces, and to different rates of change and exchange. What might follow from this? A different understanding of heritage and hence of landscape design, and designed intervention generally. I guess the heart of the project is the development of a methodology for recalibrating ground and ground events. . . (Carter, as quoted in in Gini Lee, 'The Intention to Notice: The Collection, the Tour and Ordinary Landscapes'.
>
> (Unpublished doctoral dissertation, RMIT University, 2006, 225)

38 The word 'chorography' was borrowed in the 16th century from Latin *chorographia*, which in turn comes from Greek *chōrographia*, a combination of *chōros* (place) and *graphia* (writing). This term applies equally to the process of describing or mapping a region or place, but also to the resulting description, map or graphic representation. Definition of 'Chorography' found at *Merriam-Webster.com Dictionary*, <https://www.merriam-webster.com/dictionary/chorography> [accessed 10 April 2020].

39 The Mount Noorat Project (2009) was a collaboration between Ross Gibson and Gini Lee.

40 Rebecca Solnit says: "I like walking because it is slow, and I suspect that the mind, like the feet, works at about three miles an hour. If this is so, then modern life is moving faster than the speed of thought or thoughtfulness". *Wanderlust: A History of Walking* (New York: Penguin, 2001).

41 Robert Smithson in 'A Provisional Theory of Non-sites', as quoted in 'Unpublished Writings' in *Robert Smithson: The Collected Writings*, ed. by Jack Flam, 2nd edn (Berkeley: University of California Press, 1996). See: <https://holtsmithsonfoundation.org/provisional-theory-nonsites> [Accessed 10 April 2020].

42 Susan Sontag, *The Volcano Lover* (New York: Farrar, Straus and Giroux, 1992), 24.

43 Gini Lee, on Susan Sontag, *The Volcano Lover*.

44 Cinzia Schiavini, 'Writing the Land: Horizontality, Verticality and Deep Travel' in William Least Heat-Moon, *PrairyErth* (New York: Houghton Mifflin, 1999).

45 Kalpita Bhar Paul, 'The Import of Heidegger's Philosophy into Environmental Ethics: A Review', *Ethics and the Environment* 22, no. 2 (2017), 79–98. <http://www.jstor.org/stable/10.2979/ethicsenviro.22.2.04>.

46 Arne Naess and George Sessions devised the deep ecology platform, also known as the eight points of the deep ecology movement. . . . The platform is not meant to be a rigid set of doctrinaire statements, but rather a set of discussion points, open to modification by people who broadly accept them. In fact, the version given here was modified from the original by participants attending a deep ecology course held at Schumacher College in 1995. Some deep ecology supporters regard the platform as the outline of a comprehensive ecosophy in its own right. See: Drengson, Alan, 'Some Thought on the Deep Ecology Movement', *Foundation for Deep Ecology*, <http://www.deepecology.org/deepecology.htm> [accessed 10 April 2020].

47 Stephen Harding, 'What Is Deep Ecology?' See: <https://www.schumachercollege. org.uk/learning-resources/what-is-deep-ecology>.

48 Rachel Carson, *Silent Spring* (Boston, MA: Houghton Mifflin, 1962).

49 Martin Heidegger, 'The Thing', in *What Is a Thing?*, trans. by William Baynard Barton and Vera Deutsch, 6137 (Washington, DC: Henry Regnery Co., 1969).
    The corollary here is the 'emptiness' of the jug that is in fact, 'not nothing', but the space that allows the jug to function as a jug.

50 In 2018, autumn rainfall was 57 mm below average, the driest autumn since the 1902 Federation drought. But the 2017–18 year overall was not one of the driest on record—there were 19 other years that were drier. This drought is severe but has not been as prolonged as the millennium drought—the longest dry spell in history, which saw nine consecutive years of low autumn rains, crucial for the southern cropping season. The numbers are based on the average autumn rainfall from 1961 to 1990, which was 102.3 mm. Catherine Hanrahan, 'Chart of the Day: Is This the Most Severe Drought in History?', *ABC News On-line* (6 August 2018) <https://www.abc.net.au/news/2018-08-06/how-does-the-current-drought-compare/10055414> [accessed 10 April 2020].

51 James Elkins, 'Writing Moods', in *Landscape Theory*, ed. by Rachel Ziady DeLue and James Elkins, (London: Routledge, 2008), 69 [Author's italics].

9 The slow pursuit of fidelity

Three key words resonate strangely in this title: 'fidelity', 'slow' and 'pursuit'. A 'slow pursuit', for example, seems contradictory if the usual sense of pursuit is a chase, a hunt for elusive prey, or a clever tracking or following. Equally unusual is the suggestion of 'pursuing fidelity' when the notion of faithfulness seems more of an abstract value or moral goal. Together, these three words conjure the passage of time and the long commitment of followers on a lifelong quest.

That would not be far from the truth of Alexander and Susan Maris' artistic practice and philosophical adherence to a value of faithfully seeking 'perfection'. The reality of their daily lives is played out in rural Scotland, in the highlands of western Perthshire at Kinloch Rannoch, under the mountain known as Schiehallion.[1] In the Maris household at the Old Laundry Cottage, also known as Rowan Cottage, the mountain fills the windows overlooking the back garden. And indeed, the garden itself seems to be the very beginning of the slopes at the foot of the Fairy Hill; magical territory where transformations seem entirely possible (Figure 9.1).

Susan and Alexander Maris (simply called 'Maris' by wife, family and friends) moved to the Laundry Cottage, just outside the village of Kinloch Rannoch, in 2010.[2] But in truth, their links to the place go back much further. Maris explains that Kinloch Rannoch was a place where they had always come.[3] When their son, James, was born, they came to Rannoch Moor to place his foot on the ground for the first time in the Black Wood. Their own private ritual was a celebration and a grounding, in both senses of the word. Maris continues, saying:

> It goes way, way back, our relationship with this place. But it is not just the magnetic centre, the *axis mundi* of our tiny world. It encompasses the moor, the Black Woods, Schiehallion, maybe this is the Eastern edge; between the Black Mount (mountains) and Schiehallion has been an obsession even before I left [the family home].[4]

In fact, Maris traces his long obsession with the area back as far as 1970, which also happened to be the same time when Joseph Beuys visited the

Figure 9.1 Schiehallion, as seen from the Maris' back garden at Laundry Cottage (2014). Photo: Mary Modeen.

area. This visit was at the invitation of Richard Demarco, leading arts im-presario in Scotland for the last 60 years.

It was Demarco who took Beuys on this journey in 1970 on 'the Road to the Moor of Rannoch and the Road to the Isles', to encounter the reality of the land itself. 40 years after this original visit, in a celebration to memorial-ise the original event, Demarco remembered (Figure 9.2):

> when Beuys came (to Scotland), he concentrated on the whole idea of *what is it* that is in the very stuff and substance of Scotland, in the earth, in the sky, in the light, in the summer, in the winter, in the autumn, in the springtime. *What is it* that gives you at its best the poetry associated with the Celts and the genius of St Columba who Beuys regarded as one of the great revolutionary artists.[5]

Like Beuys, Maris holds dear this historical, mystical and deeply shaping quality of the moor. Like Beuys, he and Susan feel connected to the place in ways that transcend explication, or even knowingness:

> . . . Beuys' fascination for the moor (was) somehow tangential or even identical with our fascination. Something that it seemed to him is ex-actly the same the one-word poem or two-word poem: Rannoch Moor, Kinloch Rannoch.[6]

Without stretching overly far, Maris' words are reminiscent of Heideg-ger's sense of dwelling poetically.[7] More on this connection will follow.

Figure 9.2 Richard Demarco on Rannoch Moor, 2010. Photographer unknown.

In the meantime, the Maris' response to dwelling attentively is at the core of their lives. The colours of the day, light, sun, clouds, leaves, heather, rocks, and even the subtle changes in the alpines and lichens form their daily backdrop. Susan's attention to the colour purple has been a particular pursuit. As an expert in natural fibre dyes, she has investigated many organic materials that she has concocted into batches of dyes, comparing myriad shades of purple. Her dyed wools, cottons and silks have been meticulously labelled in days, locations, temperatures, times and material source of the dye as well as the fibre source. Skeins of brownish-purples, blue-ish violets, grey-ish fuschias and vibrant burgundies hang in her back room and from most rafters in the cottage. They evoke the days that spawned these hues: winter mornings in the Black Woods; shadows in the limestone hills above the lochs and lochans; sunny August days with all the purples of heather, rosebay willowherb, thistles, devil's bit, marsh orchids and harebells: infinitely varying purples. It is *exactly* 'the colour of the day' – the insistence on this one precise colour – distilled into one rich hue, each differing from each, that Susan has discovered: she extracts the colour from lichens, water, peat and plants, allowing the alum and other additives or variations in the dyeing process to alter the saturation of these colours. Even now, she is exploring the time of collecting dyestuff source materials, collecting specimens at precisely midnight, or under the full moon or the sliver of a waning moon, or in the rain, to discover potential differences: she extracts the colour from lichens, water, peat and plants, allowing the alum and other additives or variations in the dyeing process to alter the saturation of the colours. And of course, being Scotland, there is plenty of rain to aid these experiments. Susan has found

Figure 9.3 Orchil 1, Alexander and Susan Maris, *Prima Materia: Orchil*, Silk dyed with lichen from Rannoch Moor (2013). Photo: Susan Maris.

distinct differences between time and conditions of the dye source, and can expand her annotated samples to show how stages of the moon, or time of day, or water sources, display markedly different colours. She smiles when saying this, as an alchemist of lichen, marvelling at the significance of mysterious powerful unseen forces that lie invisibly within the dyed fibres. The colours show – and don't show – the perfection of that one day's efforts, in which collecting the lichen began a process that spawned this colour, so simply, so complexly (Figures 9.3 and 9.4).

Noting colour in all its nuances is one form of attention that is repeated in the Maris' practice. This exchange between the Marises and the author evidences more than a hint of another kind of attention, in addition to their humour and their honed material sensibilities, and being ecologically attuned to the moor. Specifically, this is a kind of contemporary practice that harks back to the deepest tradition of alchemy; a transubstantiation – of water to lichens to dye to regal purple textiles. The instinctive, or intuitive comprehension that runs deeply throughout their lives, and impels them to celebrate with private rituals moments of heightened significance with the land itself, appeared in early instances. For example, Maris discusses his final year of undergraduate work at art college (Figure 9.5):

Maris: . . . it was just after Christmas, so I was going to be having my degree show that June (1981, Glasgow School of Art). I hadn't a clue what I was going to do. I had been in sculpture and I was fiddling about. . . I mean the sculptures I was making weren't rubbish, but they were just dour, unfixed. It was like learning to make things. And then, I just had

Figure 9.4 Ochil 2, Alexander and Susan Maris, *Prima Materia: Orchil*, Silk dyed with lichen from Rannoch Moor (2013). Photo: Susan Maris.

an inspiration, a genuine inspiration based on an Inuit or Eskimo kayak in the museum. And I thought, that's fantastic, and it's exactly what I need, because what I wanted to do (and I didn't know why I wanted to do it) . . . I wanted to kayak from the Black Mount, which I knew very well as a mountaineer.

But I knew that all water flowed eastwards and eventually got to Rannoch Station. I thought, well, I'll just do this for the sole purpose that I would collect a jar of water from every named river and every named loch. There happened to be six of each. So I made two boxes with six jars and waxed lids, and I made a kayak – not an Eskimo one, but it was a canvas-covered black kayak. And so that was my first absolute piece for the moor, about the moor. But I never ever made that journey. I kept putting it off and putting it off, and I realised that something wasn't going to work-wouldn't have worked. . . when things we know now wouldn't have made it possible, physically. About a year and a half later I took the kayak up to the shallows of Loch Ba and I set fire to it, and I watched it be consumed by flames.[8]

From canvas kayak to ashes, the juxtaposition of water and fire was another transformation, deeply significant to the artist. And now, in hindsight for both Maris and Susan, and that planned but travelled journey as well,

Figure 9.5 Kayak on Loch Ba. Photo: Alexander Maris.

and the route by connected waterways, all have deep significance. By way of another example, the artists live intentionally simple and quiet lives, on the land and close to its seasons and pathways. But they have as much of a respectful distance as they do a closeness to their beloved Schiehallion. Maris declared:

> the interesting thing about the relationship with place is that some-times you have to. . . it's a bit like taming a horse. You have to look at it askance. You can't just go and climb Schiehallion. We've lived here for three years and maybe only been up Schiehallion three times, you know-the whole time? But we've trodden all around it and got to know that it is not just a cone. It's convoluted with tiny valleys and waterfalls and miniature forests and plants, so longevity in a place is important.[9]

Time to know the place, time to take it in slowly, to live its various real-ities is central to the Maris' commitment to Kinloch Rannoch. They trav-erse the riverbanks and wetlands, the banks of Loch Rannoch and the high moors daily. One of their special places is a clearing in the Black Wood (Figure 9.10), the site of a former croft, long uninhabited now, but discern-ible through the ruins of the walls, the rowan planted at the edge of the ruins, and the raised bed for what would have been the drained land for kitchen garden and dooryard. Time impinges multiply in this special place for them: not only has it become the site of accrued memories of visits and small rituals (the touching of their son James' first foot on the ground). But it is tangibly the spectral memory of the past inhabitants and their toil at place-making, the stewardship of the land still evident in the trees of the forest and the drills of past plantings. Maris and Susan can name every bird-call, can compare the effects of several winters, and note the current long, slow decay of stumps from trees cut for the Second World War on the forest's perimeter. This is the knowledge of deep-mapping, the lived generational experience of attentiveness that layers past histories, human events, non-human activities and eras of growth and decay. It is also the basis of all science and art as well, founded as they are on observational attentiveness.

Indigenous peoples around the world share this type of deep respect for the people of the land, 'local knowledge' as some call it, that stands as the foundation for holistic and sustainable ecosystems, for entire belief systems. This terminology is one of contemporaneous construction, but could equally be called the constructed narrative(s) of place, or Dreamtime, or vibrant ecologies.

'Vibrant Matter' is the name that Jane Bennett, in her book of the same name, applies to the potentiality of all matter to interact complexly. 'Vital materialism' is "a materialism in which matter is an active principle and, though it inhabits us and our inventions, also acts as an outside or alien power".[10] In this vitality, we find the first evocation of fidelity (Figure 9.6).

Figure 9.6 The Pursuit of Fidelity. 15th C. German tapestry: the text translates as 'We are hunting for fidelity and if we find it we would rather live in no dearer time'. Photo: Maris, © Glasgow Museums.

Fidelity is a concept that implies an idealised quality: some idea, person or thing to which one can display faithfulness or loyalty. The ideal for Alexander and Susan Maris is seeing, understanding and achieving the balance of material presence, and its harmony with its own potentiality. In an interview that has been transcribed, the decision was made to use three different fonts to reflect the three different voices: Maris in regular text; Susan in italics and the author asking questions in boldface. To convey our three voices in discussion, in this example Maris tries to offer an example of his long-standing aims, saying[11]:

MARIS: I tend to think I was always, *I* was always, personally, trying to make nothing. I was looking forward to the day when I would not have to make anything. Because it would be . . . Everything, I think, can be summed up by the tapestry, *The Pursuit of Fidelity*. The lovers on horseback chasing the stag, which represents fidelity. And I interpret the net and the hurdle that they are chasing the stag into, the thing that they desire most which they cannot catch, has had to be, they'd had to use intellect and anticipation to catch it, so they've had to conceive of something, so the concept of the net, and the net's a very interesting thing, because the net has to be invisible to the stag. It has to be so ephemeral that it cannot be seen, but it has to be strong enough in order to capture the quarry. . .

The Marises use their creative practice – an inseparable component of their daily lives as they live simply in the highlands – precisely to achieve their fidelity to a kind of perfection. This perfection conjoins daily routines in balance with the environment, with growing or finding food, school for their son, making art, earning enough money to get by, stewardship of the land, close observation, joining the local community, caring for small things, and generally living their beliefs in a careful and harmonious balance with their environment. The artworks that emerge in this slow practice come as a kind of physical/metaphysical meditation in this holistic process. Sometimes the artworks develop as a kind of residual material legacy of transitional acts, like Susan's purple dyed silk shawls, for example, or her woven dyed tapestries. Or sometimes they are photographs, taken as a document of fleeting moments, or commissioned films, made to capture slow observation and implied poetic associations.

If we, in our contemporary times, have failed to reach this notion of balance, perhaps our failure has lain in the lack of understanding how deeply interconnected, how truly deep and wide, are the links in all matter, and how our human actions in brief lives have long consequences beyond our ken. Taking time to abide with a place, to open up to the long lessons of it, means creating the space for things to happen. Given time to dwell, creatively, without rushing, is to take in the many manifestations of how interconnections mean the agency of all matter, all life, all things.

Heideggerian notions of dwelling, for example, implicitly demand slow attention to the world. In these terms, slow means not just achieving or attaining, but more importantly, staying with, abiding – and hence, dwelling reflectively.[12]

MODEEN: . . . **You're using kind of a Beuysian frame to put over this narrative, but in a sense, you're also having your own narrative by being here, by living here. You're, in a sense, replicating – not replicating Beuys, but . . . re-establishing a different kind of framework that will carry on with another kind of frame that other people can see. So the burying of the wooden object, or the stone . . . the ways that you signify, are going to be another framework that's part of this on-going series of narratives that are quintessentially cited narratives, aren't they?**

MARIS: Yes, because they're also personal. So, we keep coming back to this term, meta-biography, they're meta-biographies of our existences, because we all have many existences, and just a frame of consciousness that we happen to be in, in our teens, twenties, thirties, and so on. And it's like we've set ourselves with clues, buried clues, for our future selves. And we didn't know it at the time, that we would need those things. So that's why we consume them, we . . . our previous selves maybe had a subconscious knowledge that that was going to be required. The stone fish was acquired in a very dark period in our history. And we had just got a bursary to go photograph bits of the old forest. And I said to Sue, I will take you to the Island and I will . . . one day we'll go there because the forest is virginal there, it's unchanged. And I was nervous about it. We got a boat, piled all the stuff in, went there, and of course, in researching it, I did go in my old notebooks to see what I'd written in there and so on the last page of the episode there, was mention of the stone fish, and I thought, and so it came back to me and I thought, crikey, I wonder if it's still there. So as soon as we landed, I went to see if it was still there. And the hollow trunk had rotted away. Twenty years, it'd been there. You know the winters on Rannoch Moor: minus fifteen is normal. This tiny, delicate slate fish, the hazelnut tree was grown up, and perched, literally, in a branch like that, on a mossy cushion, was just sitting there on the bark. It completely shocked me, to the core of my soul. When I picked it up, it was unblemished, there was not a mark on it, moss molecule attached to it. It was astonishing. And I knew it meant something, but I didn't know what it meant. You never know what these things are about. But it did.[13]

Can objects mean something, in themselves? A resounding and unequivocal 'yes' would be the answer of Maris and Susan . . . and perhaps most artists, whose approach to the phenomenological world begins with all matter. But, like Heidegger articulated, to 'dwell reflectively' is to engage in a constant process of reflecting on the 'how' and 'what' of things, and to set this aside and start again.[14]

Dwelling means always emptying in order to fill again with potential. To stop seeing is to forget to empty, to renew, in order to see afresh. This freshness is the renewed encounter, whether one is doing something repeated continually, or seeing a thing that has been set aside for many years, like Alex's stone fish encountered again.

For these two artists, both Maris and Susan, part of the attraction of Beuys' actions and creative work was his philosophical stance, which they shared with an intuitive and deep appreciation. This belief system may be characterised as the same as Beuys' conviction that all 'substance' (or matter) has *both* physical and spiritual qualities. Victoria Walters considers this in her study of Beuys' Celtic works:

> In many respects, the figure of the shaman represented, for Beuys, forgotten human powers that needed to be rekindled not only in order for social, revolutionary change to occur, but to confront an imminent human and environmental crisis . . . Through engaging with his understanding of both the Celtic and pre-Celtic world and the role of the shaman in the history of human consciousness, Beuys sought to expand language itself and to investigate the role of language in all sculptural processes, as a means to stimulate people's perception of and engagement with the shaping of both visible and invisible worlds.[15]

Shamanism as defined in this sense has as much of an ability to 'stimulate people's perceptions' as it does to conjure or transform. One could make a case for the Marises as a kind of next generation of artist-shamans, then, if their work has the capacity to connect the physical and spiritual, and thereby to stimulate perceptions. On the other hand,

MARIS: . . . I was trying to come up with the perfect artwork, intellectually speaking, something absolutely definitive and perfect and it had to have all the philosophy about art and objectivity wrapped into one thing. And that's when I came to see Sue, because she was studying philosophy. And as quick as Sue was reading philosophy books, I was burning them, ripping them into paintings.

MODEEN: **This is a lot like Beuys' fat. You're burning her (Susan's) books?**

MARIS: I was burning her books, She'd said. . . sums. . . What was it?

SUSAN: *Sums! That was disturbing because I would be a quarter of the way through the book and I'd realize that on the inside of the cover, there was what dimensions of the canvas it was going to make and therefore I knew I'd have to be hasty about it.*

MODEEN: **Finish the book?**

MARIS: We burned, read and burned, lots of philosophy. Writing about colour, too.

SUSAN: *Unread, too, yes, it was nice.*

MODEEN: **Did you keep the ashes (for making artworks)?**

SUSAN: *Yes. There's a nice distinction about the unread and read books, as well, the colour was quite different. It was quite beautiful.*

MARIS: Yes, well, we eventually landed on Jacques Derrida's *The Truth in Painting* as the definitive novel to burn.

MODEEN: **Best burning material? 'Words to burn'?**

MARIS: This is the canvas [see Figure 9.7]. And this is the colour it produces, this beautiful grey colour. And this dark grey, if you don't read it. And light grey if you do. . . It's amazing. It starts to oxidize as soon as you start to open a book, crack the seal, open it. And now the content begins to evaporate.

MODEEN: **So the darkness is all the kind of thought material that's in there before it is consumed. . .**[16]

The Marises are able to embrace unresolved ambiguities with happy abandon. They choose to create a space for contemplation as a creative act in itself, living with and within a rural highland area. Eschewing many of the trappings of cosmopolitan life, and the hectic pace as well, 'Maris X 3' – as they like to refer to themselves, inclusive of son James – they have adopted the slower pace, closer knit communities and wider open spaces of remote rural life instead. This is not a negative choice; theirs echoes the choices of,

Figure 9.7 Alexander and Susan Maris, *Extracts from the TRUTH IN PAINTING* (1990–93, revised 2006), acrylic paint canvas and ashes from one read and one unread copy of Derrida, *The Truth in Painting*. Photo: Maris, exhibited at the Ingleby Gallery, Edinburgh.

Figure 9.8 Theraputicum: Susan Maris, fishing. Photo: Maris.

say, Thoreau nearly 160 years earlier, as he states in *Walden*: "I went to the woods because I wished to live deliberately. . ."[17]

To live deliberately, and to deliberate is to ponder, to reflect, to take the time to think more fully and act with this mindfulness. It is another iteration of the slowness and fidelity in the chapter's title. For these two artists, this plays out in daily actions that include gardening, walks in the woods, foraging, fishing for food for the table and in a myriad of small actions that are repeated in rural communities across the highlands and islands. Most inhabitants of the highlands and islands have more than one job; all have the many-sided tasks associated with living in a climate that demands changeable activities as the seasons change (Figure 9.8).

Implicit in this lifestyle (or should one say, implicit with this philosophical foundation of belief?) is the ability to accept conditions that arise (limited broadband, patchy mobile communications, unseasonable frosts or a month of rain) and an attitude of mindful watchfulness. The yellowing leaves on a patch of ground that shows the presence of biotoxins, the spoor of pine martens, known almost as closely as family members, shows well-trodden paths, or the growing population of sparrows presaging a milder winter. . . all are meaningful and significant. Close observation in this 'slow residency' reveals a hitherto unexpected wealth of interrelated connections, each interacting with other elements. 'Reading' the land is a process of gathering information in unfolding metaphysics, as well as physical inter-relationships and mutual interactions. Applying a hermeneutics of ecology, it signifies across living and non-living, human and non-human animals, in ways that intersect and resonate. Witnessing and reflecting are the Maris' elected roles. Self-reliance, careful attention, moments of stillness and creative actions are what is required to achieve this (Figure 9.9).

Figure 9.9 Susan Maris, works-in-progress with natural fibre dyes from lichens, fungus, wood and berries. Photo: Maris and Susan Maris (2014).

Perplexity itself is nothing to avoid; with a laugh or a smile, Maris will shake his head at the legends of cross-bill nuthatches or Sue will wonder how fungus collected in a new moon might differ from that collected at full moon in its ability to affect the intensity of purple dyes. One is reminded of the profound knowledge, the 'local lore' of indigenous peoples, and is aware once again (as if anyone needs reminding) of the value of long and sustained observation, particular to a specific place (Figure 9.10).[18]

'Practice' for the Marises is synonymous with understanding deeply what it means 'living with place, living in place'; their performative actions mean 'staying with' – *biding with*, in Scots vernacular – their chosen home. These prepositions – *with* and *in* – and *biding* are all part of the wonderfully subtle Scottish attention to relationships with place itself. In fact, it is a little recognised but deeply significant aspect of Scottish character which features in the finely attuned nuancing of prepositional status: 'outwith', for example, or 'hereby'.[19]

These are not anachronistic terms. They are in active use and spoken by young as well as old, and mark a distinctive relational sensibility in long term residents. These words betray a linguistic spectral trace of positional relationships as well as an underlying sense of cultural value and philosophical importance to make these interactions clear (Figure 9.11).[20]

These values are evident in most of Alex and Sue's artworks. For example, in their film *Uriel*, the rhythms of the film echo a monastic schedule. Conceived as a kind of filmic remake of a medieval book of hours, the tempos

Figure 9.10 The Black Wood, by Loch Rannoch, in April. Photo: Mary Modeen.

Figure 9.11 Still from *Uriel*, film by Susan and Alexander Maris (2008); duration 34
    min: 8 sec.

of eight sections are defined by light. Changeable qualities of sunlight and
clouds, passing rain and darkening skies, stand in relation to each other as
visual meditations on passing time and shifting patterns in nature. To a background of clinking bells from a distant herd of cows, views of a mountain

slide slowly across the screen, majestic, terrifying, stunning. Great restraint was exercised in the edit of this work: the land is allowed to speak for itself. Viewers make what they will out of watching light move across a rocky façade of the mountain in this film; direct observation through the eye of the camera is done discreetly by the artists without an imposition of overt signification, other than the frame of the lens itself. It evokes Maris words, cited earlier "*I* was always, personally, trying to make nothing . . . Because it would be 'perfect' . . . fidelity . . . a balancing act between not quite existing, only existing just enough to do what it wants to do".[21]

In the overlay between fidelity and 'perfection', as Maris describes it, is the 'balancing act between not quite existing' and being present with actions that touch so lightly as to be unobtrusive in a natural world. In *The Pursuit of Fidelity* (Figure 9.6), every part of the visual narrative encapsulates potential: the lovers embody the potential that between themselves synergistically makes something more than two. The deer signals the potential of animal life, of animated vitality in the forest and as food for the hunters. And the net, read as a planned intervention, is the signifier for human intention and action, an example of agency in the natural environment, which only works when it is seen as 'barely existing', when it mimics nature most closely.

These three words still chime as bells: fidelity, pursuit and slow. 'Fidelity' is usually used in relation to 'fidelity *to*' something: a faithfulness to a person or a cause; a faithful reproduction of sound qualities; a faithful reproduction as a physical likeness to the sitter in a portrait.[22] Fidelity in the sense that applies most closely to the Maris' practice might be said to be a faithfulness to a whole lifetime of observed nature, of the determination to seek nature's secret rhythms and materials, and to work with these to bring a heightened sense of awareness to some aspect of this vibrant material potential and ontological interaction. Lest that sound rather grand, Maris refers to this as his 'almost nothing'. But their practice is not nothing. It betrays a restrained hand, a refusal to over-inscribe too insistent a narrative or mythic structure, and instead demonstrates a willingness to let the content unfold slowly, of its own accord. The structure it does have lies invisibly in the artists' conceptual framing, their deep knowledge of mythology (especially Celtic myth), of philosophy and of contemporary art practices. Their material knowledge is always present, and exacting in its control (Figure 9.12).

Vital materialism is relevant to this discussion, perhaps not because the artists rely upon promoting this thinking, but because they enact it, knowingly or not.[23] Part of this vital materialism is concerned with a breaking down of hierarchical structures of being, especially a schema in which humans have sole agency as actants, and prime status as unquestioned masters of their world. Instead, vital materialism in its many forms sees matter (animal, vegetal and mineral) as a kind of common denominator that connects all living and non-living things.

Figure 9.12 Lochan on the Moorlands. Photo: Mary Modeen.

> It is a feature of our world that we can and do distinguish . . . things
> from persons. But the sort of world we live in makes it constantly possi-
> ble for these two sets of kinds to exchange properties.[24]

'Thing-power', as Bennett refers to this attention to ubiquitous and shared
materiality, is a revaluing of the ability of matter to, well, 'matter'. The point
here is that if the component parts of animals – human and non-human – and
every aspect of the world they inhabit are building blocks made from one
and the same, then it is the agency of each object/subject (which must be
ultimately indistinguishable) that displays the 'vibrancy' of potential and
interactions.[25] Barad insists that "matter and meaning are not separate entities".[26] Bennett agrees, and says she is

> trying to raise the volume on the vitality of materiality *per se*, pursuing
> this task so far by focusing on nonhuman bodies, by, that is, depicting
> them as actants rather than as objects. But the case for matter as ac-
> tive needs also to readjust the status of human actants: not by denying
> humanity's awesome, awful powers, but by presenting these powers as
> evidence of our own constitution as vital materiality. In other words,
> human power is itself a kind of thing-power. At one level this claim is
> uncontroversial: it is easy to acknowledge that humans are composed of
> various material parts (the minerality of our bones, or the metal of our
> blood, or the electricity of our neurons). But it is more challenging to
> conceive of these materials as lively and self-organising, rather than as
> passive, or mechanical means under the direction of something nonma-
> terial, that is, an active soul or mind.[27]

Poetic associations in the 'geopoetics' of this book's title imply a constant folding and unfolding. The poet folds her own lived experience and observation into language that encodes this 'being-in-the-world'. The reader reads and in so doing unfolds his own understanding, partly through the convention of decoding language, but also partly through unfolding his own lived experiences of the past as they touch the present, and unfolding the present as it opens out in many futures. If one adheres to the belief that all matter is capable of vibrant interaction, as Bennett would have it, then this being-in-the-world is a sharing of potentiality with all other matter, living and non-living.[28]

The poet Mark Strand has captured much of this in his poem, 'Keeping Things Whole':

> In a field
> I am the absence of field.
> This is
> always the case. Wherever I am
> I am what is missing.
> When I walk I part the air and always
> the air moves in to fill the spaces
> where my body's been.
> We all have reasons for moving.
> I move
> to keep things whole.[29]

Rosi Braidotti, philosopher and feminist scholar, builds a case for the importance of the effects of this kind of new materialist influence in seeing beyond the barriers of outdated models of understanding. She writes:

> the conditions for renewed political and ethical agency cannot be drawn from the immediate context or the current state of the terrain. They have to be generated affirmatively and creatively by efforts geared to creating possible futures . . . and by actualising them in daily practices of interconnections with others.[30]

It is this ability to see interconnections, to act ethically within these vibrant interactions, and to resist categories or structures that divide artificially (with the unfortunate effect of removing the responsibilities for ethical actions) that she emphasises. She continues by saying:

> Prophetic or visionary minds are thinkers of the future. The future as an active object of desire propels us forth of a continuous present that calls for resistance. The yearning for sustainable futures can construct a liveable present . . . The future is the virtual unfolding of the affirmative

aspect of the present, which honours our obligations to the generations to come.[31]

Unfolding, reading the land and creative artistic responses to the natural environment brings us full circle in this discussion, back to the Maris' practice, circling back to the reiteration of poetic association, and unfolding. If poetic association works as a hermeneutic of unfolding, that which is revealed in the space that opens out is a rich combination of recall, linking memory, matter and the embodied experience of life, with the present. Poetics as applied to the environment must be as rich in its invitation: geopoetics is a return to the response we feel for the earth as the first mother that has given us life. The geopoetic project is not one more contribution to the cultural variety show, nor is it a literary school, nor is it concerned with poetry as an art of intimacy. It is a major movement involving the very foundations of human life on earth.[32]

It is on this scale that the magnitude of ideas, the politics of a profoundly different reorganisation of the world (and the urgency in timely actions), is emerging in philosophy and is finding a resonance in art and artistic practices. Like many artists sharing these sensibilities, Alexander and Susan Maris have forged a collaborative practice that seeks to promote attention to the interconnectedness of all matter, to work in tune with nature and natural rhythms, and to structure the most sensitive – and restrained – of artistic productions that capture, consolidate and open out imaginative visions of life as lived with deliberate fidelity and slow reflection.

Notes

1 In Gaelic the mountain is *Sìdh Chailleann*, known as the 'Fairy Hill 'of the Caledonians.
2 Since first drafting this chapter, the Marises have moved home and now live at Chemical Cottage, not far from Laundry Cottage. It seems somehow in keeping with their transformative art, and the name 'Alchemical' Cottage might be closer to the mark.
3 Interview by Mary Modeen with Alexander and Susan Maris, July 2014 at Laundry Cottage, Kinloch Rannoch. Hereinafter referred to as 'Interview'.
4 Interview.
5 David Gibson, 'Joseph Beuys 40th anniversary journey "to The Moor of Rannoch and the Road to the Isles"', *Studio International* (8 May 2010), <http://www.studiointernational.com/index.php/joseph-beuys-40th-anniversary-journey-to-the-moor-of-rannoch-and-the-road-to-the-isles> [accessed 21 June 2010].
6 Interview.
7 Martin Heidegger, 'Poetically Man Dwells', in *Poetry, Language, Thought*, trans. by Albert Hofstadter (London: Harper Collins, 2001).
8 Interview.
9 Interview.
10 Jane Bennett, as quoted in Diana Coole and Samantha Frost, eds, *New Materialisms: Ontology, Agency and Politics* (Durham, NC: Duke University Press, 2010), 47.

11 This transcription was quoted verbatim, with an attempt to preserve the rhythm of speech and the nature of the speakers' voices.

12 Martin Heidegger, 'Building, Dwelling, Thinking', in *Poetry, Language, Thought*, trans. by Albert Hofstadter (London: Haper Collins, 2001).

13 Interview.

14 This emphasis on materialism ('New Materialism', 'Object Oriented Ontologies', or 'Speculative Philosophy' as branded in the last ten years, is a 21st C 'turn' to a rethinking of materialist metaphysics).

15 Victoria Walters, *Joseph Beuys and the Celtic Wor(l)d: A Language of Healing* (London: Lit Verlag, 2012).

16 Alexander and Susan Maris, exhibited at the Ingleby Gallery, *The Truth in Painting* (Installation B), 1990–93 (revised 2006). Acrylic medium on canvas.

17 Henry David Thoreau, *Walden: Or, Life in the Woods*, 1854 (Princeton, NJ: Princeton University Press, 2004), 154. The full quote is:

> I went to the woods because I wished to live deliberately, to front only the essential facts of life, and see if I could not learn what it had to teach, and not, when I came to die, discover that I had not lived. I did not wish to live what was not life, living is so dear; nor did I wish to practise resignation, unless it was quite necessary. I wanted to live deep and suck out all the marrow of life, to live so sturdily and Spartan-like as to put to rout all that was not life, to cut a broad swath and shave close, to drive life into a corner, and reduce it to its lowest terms.

18 This is a pattern that is linked inextricably to the traditions of Indigenous peoples the world over. For example, traditional Māoris welcome visitors to the community with gatherings, held at their special community halls (*maraes*), by a procession (*whakaeke*) led by women who call plaintively in a ritual to address the dead (*karanga*) while leading the group slowly to the entrance of the *marae*. (Tania Ka'ai, and others, *Ki Te Whaio: An Introduction to Māori Culture and Society* (Auckland: Pearson, 2004), 75–79).

19 'Outwith' means 'in a context that stands outside the central organization or principle'; 'hereby' means 'with this here' or figuratively, 'by these means'.

20 This film was commissioned by LABoral Centro de Arté, Gijón and produced by Film and Video Umbrella. The film has been described as 'a circumnavigation of Picu Uriellu in the Picos de Europa (that) reveals the different facets, and topographical profiles, of this landmark mountain in a series of locked-off views, taken at various stopping-off points during the course of an imaginary day. Structured in eight 'chapters', which roughly correspond to the canonical 'hours' of the early Christian/medieval church, and also allude to the eight fire-festivals of the ancient Celtic calendar, 'Uriel' marks the passage of time, from an ethereal mountain dawn to the 'magic hour' before twilight, in a procession of memorable images, contrasting the elemental imprint of the mountain landscape with the fleeting presence of human subjectivity.'

21 Interview.

22 High fidelity ('hifi') by common association became the shorthand name, and now synonymous term, for stereo equipment and vinyl records that could play the most faithful of reproduced sounds. Now, wifi is 'wireless fidelity'.

23 Or 'New Materialism', as it is also called, and is an emergent mode of critical understanding that crosses many disciplines, including philosophy, geography, politics, psychology, feminist studies, ecology and the arts. It is also linked with 'vibrant matter' (Bennett); 'agential realism' (Barad); 'morphogenesis' (De Landa); 'nomadic theory' (Braidotti); and 'Speculative materialities' (Meillassoux), among others.

24 Karen Barad, *Meeting the Universe Halfway: Quantum Physics and the Entanglement of Matter and Meaning* (Durham, NC: Duke University Press, 2007), paraphrased from Chapter 2.
25 Jane Bennett, *Vibrant Matter: A Political Ecology of Things* (Durham, NC: Duke University Press, 2010), 10.
26 Barad, *Meeting the Universe Halfway*.
27 Bennett, *Vibrant Matter*, 10.
28 Ibid.
29 "Keeping Things Whole" from *SELECTED POEMS* by Mark Strand, copyright © 1979, 1980 by Mark Strand. Used by permission of Alfred A. Knopf, an imprint of the Knopf Doubleday Publishing Group, a division of Penguin Random House LLC. All rights reserved.
30 Rick Dolphijn and Iris Van Der Tuin, *New Materialism: Interviews and Cartographies* (Ann Arbor: University of Michigan Press, 2012), 35–36.
31 Dolphijn and Van der Tuin, *New Materialism*, 36.
32 As quoted in the inaugural text of the International Institute of Geopoetics, in 'The Scottish Centre for Geopoetics', and affiliated to the International Institute of Geopoetics, founded by Kenneth White in 1989. The reservations that the authors have in relation to the original stance is outlined in Chapter 1.

10 'Fieldwork', reconsidered

In the Preface to Edward Casey's important book *The Fate of Place*, he aims to set out a firm etymological link between the original words signifying 'politics' and 'ethics', which he writes, "go back to Greek words that signify place: *'polis'* and *'ēthea,'* 'city-state' and 'habitats'", respectively.[1]

While having an intellectual empathy for the desire to affix community and ethics to specific places, Casey in this instance mistakenly translates *ēthea* as 'habitats'. In fact, it translates more accurately as 'habits', closer in meaning to *'ethos:'* how one is raised, customs, shared values in upbringing. But despite the slip in translation, he takes an indirect route to where he is intending to go, as he continues:

> . . . More than the history of words is at issue here. Almost every major ethical and political thinker of the century has been concerned, directly or indirectly, with the question of community. As Victor Turner has emphasized, a *communitas* is not just a matter of banding together but of bonding together through rituals that actively communalize people – and that requires particular places in which to be enacted.[2]

The emphasis here is on community *(communitas)*. Writing at the end of the 20th century one of the most salient elements of Casey's work, and that of a great many other theorists, was the overlapping tensions between community as a social construct and the sites for this place of coming together. That primal tension has not been resolved in any real context. Rather, looking back two decades on, we in the 21st century see the tensions underlined, if anything, with even more critical urgency. The place of each human is not a given; in contrast to any previous assumptions that the right of any human by virtue of being born, to exist *in place* is not an automatic right to which we humans as a species in this world still adhere. It only takes the most cursory glance at the daily news to find instances across the planet where an inhabitant population is displaced, evicted, moved out, or destroyed. The sad and critical lessons of the 20th century have not been learned, in large part for reasons Amitav Ghosh admirably sets out in his *The Great Derangement: Climate Change and the Unthinkable*.[3]

That critical analysis is not the place to end this book, however. In the two decades of the 21st century to date we see that there is an even more stringent insistence on place, or in the parlance of the academy, 'a spatial turn'. And one thing we share as a *leitmotif* throughout this book, is that there is still the possibility of a 'reimagined fieldwork' for practitioners of all sorts, a reinvigorated *communitas*, as there is still a possibility of finding examples of new ways to address place-based praxis. The spatial turn is one that makes complete sense and has helped bring us to the new politics that Bruno Latour calls that of the Terrestrial.[4]

The Spatial Turn

Scholars have been increasingly concerned with what maybe loosely called 'the Spatial Turn' since the 1970s and 1980s.[5] Combinations of studies in various fields, such as the emergence of phenomenology in Philosophy, Bachelard's *Poetics of Space*, and the work of Lefebvre in *The Production of Space* set out different ways of conceptualising space; the attention to cultural differences and re-indigenising knowledge has risen to the fore in Indigenous Studies; in art, spatial concepts have shifted from public art to installations to peripatetic and performative walking practices; and cultural geography in an emergent mode has featured recent research that merges human, non-human and geophysical mapping. Pressures of economic sustainability and damage to ecosystems, coupled with concerns for social justice and conservationists' urgent pleas for halting the decimation of species, makes clear the emphasis on increasingly contested land usage and its consequences. These concerns have brought disparate people – scholars, activists, indigenes, policy-makers and members of grassroots communities – together in an effort to pay closer attention to issues of space and place. The 'turn', as Jo Gullie suggests, is the backward glance from where we have come, and the realisation that the last decades have involved a series of disastrous choices, particularly by the more privileged members of the human species.[6] All this notwithstanding, we need to listen when Donna Haraway urges us to 'stay with the trouble'.[7] If this 'spatial turn' as a form of reflection and analysis does anything useful, it reconsiders what has gone wrong, what has been missed, and suggests how new actions might be taken that go some distance in rebalancing and reconnecting life styles to our environment. Staying with the trouble in the manner we uphold is a commitment to being *in* the process of correcting harm and regenerating sustainable alternatives, as well as articulating all the reasons why this should be so.

The preceding nine chapters have considered *ecosophy* – that is, the philosophy of sustainable and balanced ecosystems – as a prelude to any application of life economies; *deep-mapping* as a means of capturing information that exceeds any single discipline, assumption or perspective; art praxis, creative engagements and ecological actions, taken as determined steps to live deliberately and in balance with the environment; and 'slow residencies', which are self-styled commitments undertaken by individuals to the longer timeframes of commitment to action. In citing the various theoretical, practical and historical influences in the thinking that underlies these concerns, we co-authors have joined in collaborative discourse, actions, exhibitions, site visits, symposia, exchanges and funded research projects to share, debate and test our ideas in action. As every student will know, learning something by reading or being told about it is one thing, and putting one's learning into practice by doing and teaching it to someone else, is entirely another. The passive learning involved in the first instance is then transformed by first-hand experience and becomes practical knowledge, embodied and enacted, in the second, active example. Each instance of active learning is slightly different when the place and other circumstances in which it occurs are themselves factored into the account. In advocating an inclusive approach to art within a polyvocal conception of an ensemble understanding of creativity we are proposing, with Anselm Franke, a new "ability to challenge monolithic modernist narratives" and prohibitions inherent in a universalist ontology, so as to create a genuine openness to "possible transformation".[8] This fundamental rethinking of the field traditionally referred to as 'art' is necessary to overcome the dominant 'colonising form of ego consciousness' that has assumed the 'singleness or unity' that underwrites 'disciplinary' thinking. As an educational orientation, this alternative approach aims to make space for a 'multiplicity of psychic voices', including those internalised 'voices that go unheeded by dominant cultural force'.[9] All this, in turn, requires approaches that move away from the overemphasis on heroic notions of action and activism that serve to marginalise or silence the multiple, and often contradictory, voices of a life-world understood as a *pluriverse* by subjecting it to a unilateral perspective.

It is this heroic fantasy that helps to underpin the *mentalité* on which the culture of possessive individualism rests, a vision of the hero as sovereign individual that internalises the attributes of "the theological God of monotheism, anomic, transcendent, omniscient, omnipotent", and is reflected in "the splendid isolation of the colonial administrator, the captain of industry", and the 'academic in his ivory tower'.[10] As an alternative to this heroic, logocentric mono-minded individualism we propose 'individuals-in-community',[11] individuals who are "constituted in and through their attachments, connections and relationships"[12] (Figure 10.1). In this respect our concerns may be said to parallel those of persons committed to a: "decolonial political vision of a world in in which many worlds would coexist", to 'the pluriverse'.[13] Our emphasis on place, then, relates directly to the maintenance of these many worlds; to a 'keeping alive' of 'local memory and imagination as a reservoir of meanings, truths, and possibilities for a different future'.[14]

Citing novelist Carolyn See, Dear quotes her dictate to writing students:

When I talk to my students about writing, I talk about characters and plot, space and time, but, most important, geography. Place has always

Figure 10.1 Performance walk as part of *Invisible Scotland*, international five-day conference, 2013. Photo: Iain Biggs.

been where you start. The more specific you can get about a particular place, the better chance you have of making it be universal and of really grounding it.[15]

As See understands, the setting of the writing is akin to the first understanding we have of our own contexts; we are *who* we are because of *where* we are.[16] For her and so many other writers, the awareness of particularities of the encompassing world in which we're enmeshed amounts to a continuous choreography of movements in, with and around the spaces we inhabit. It is philosophers, physicists, architects and geographers who have theorised more than a simple occupation of space as we have seen in previous chapters, and as such, in several forms. The nature porosity, of permeability in living beings, suggests that we are co-present with all other matter and organisms in our time-place locus of being.

Pursuing this idea of permeable space, moving beyond any possibility of closed containment, whether located within a single living organism or as a larger construct of assembled organisms, such as in a city, it is worthwhile to investigate further this notion of 'porosity'. Walter Benjamin co-authored an essay with Asja Lacsis in 1925, which suggested that the way that we know things often relies upon a language or visualisation of containment. In other words, he began by discussing, and then moving away from, closed forms. Experiencing a porous city, he said in this essay about Naples, for example, one walks through neighbourhoods where there is an identifiable character, and we note its flavour or distinctiveness in making sense of where we are. But there are 'spatial porosities' he suggests, and these are instances where there is an overlapping or permeation of barriers and boundaries.[17] In fact, in these instances, it is difficult to determine boundaries at all, and instead one must experientially note qualities rather by *transitions* instead of boundary lines.[18] Commenting on this perception, Victor Burgin highlights[19]:

> the pre-Oedipal, maternal, space: the space, perhaps, that Benjamin and Lacis momentarily refound in Naples. In this space it is not simply that the boundaries are 'porous', but the subject itself is soluble. This space is the source of bliss and terror, of the 'oceanic' feeling, and of the feeling of coming apart . . .

In this excerpt from the essay *Neaples* (1925), Benjamin emphasises the qualities of 'barrierlessness', or porosity (*porositat*) that allows for reinscription, palimpsest and qualitative multiplicities. Another word he used to describe this effect was *durchdringen* (literally to 'push through'). His sense of simultaneous qualities prompted him to see the overlay of culture and nature as one dense place; as Gilloch notes:

> Benjamin aims to reveal the process of construction as the production of instant ruins. Naples is the perpetual ruin, the home of nothing new. In the ruin, the cultural merges into the natural landscape, beckoning indistinguishable from it. The product of instant ruins![20]

To re-emphasise this embodied sense of space and the prevalence of permeable boundaries, philosopher Luce Irigaray has written extensively on the porosity, indeed the multiplicity, of our conception of embodied place.[21] Additionally, much of the contemporary emphasis on object-orientated ontologies and new materialism, with its self-organising properties and shifting interrelated flux of being, relative to time and place, means that definition itself – in short, *any* notion of fixity – is suspect. As a result, another way to think about place is in the terms set out by Geraldine Finn as: the *space between* experience and expression, reality and representation, existence and essence; the concrete, fertile, pre-thematic and anarchic space *where we actually live*. There is a fundamental resonance here with the work of LeFebvre; in his *Production of Space*, LeFebvre discerns distinct differences between the everyday life of 'perceived space' (*le perçu*) and the constructed, conceptualised life (*le conçu*) of estate agents, cartographers and urban planners, for example. Moreover, the person who is fully human and alive (being-as-becoming) also dwells in the 'lived space' (*le veçu*) of the imagination, which includes dreams, and is kept vital by the arts and the imagination. This third, 'lived' space (*le veçu*) not only rises above the other two levels in a hierarchy of spatial awareness, but has the ability to reconfigure the relationship of the popular 'perceived space' and official 'conceived space'. This is an important addition to Finn's insight above that builds on in-betweenness. LeFebvre insists upon this power of the imagination to produce and shift spatial understandings, in effect, to *create* a new space through creative imagining. For him, it is only the complete human (*l'homme totale*) that can bring this fullness of dwelling in the synthesis of the perceived, the constructed and the imagined.

If we can accept this production of new space as suggested by both Finn and LeFebvre, we can be said to reside *within* this shared space as much as we do *beside* each other. Nor is this space simply something to be found in the writing of philosophers and theorists. It is very much the topic of so many of the works of the painter and printmaker Ken Kiff, about which he writes, in a letter about the question of 'finishing' or 'leaving' a work:

> For the work to be left, a totally unintelligible new thing has to be sensed as complete, perhaps necessitating a determination to scrap all ideas of completeness. After all, all ideas of completeness will be useless anyway.[22]

Co-inhabiting our environment, understanding the de-centred, non-enclosed, unfinished nature of our own physical beings, also invariably de-centres us as 'entities'. We *are* the places we inhabit; we are not separate from, but intrinsically a part of, the intermeshed web of matter and energies of which we are a constituent part. Reading takes up the term 'singularity', for example, as embedded within Deleuze and Guatarri's *Thousand Plateaus*, to "indicate that there is no longer a subject-position available to function as the site of the conscious synthesis of sense-impressions". The locus of self, as in so many other ways echoed throughout this book, places the subjective observer as a fragmented shifting, 'collective grouping of selves', mycelial in effect, demands the inclusion of its specific setting as constitutive of its entity.

Given this diffusion of identity outwards, from a disappearing 'internal' core to a splintering of shared atoms and microcosms with the rest of the environment, it is no longer a question of whether it is necessary to conduct fieldwork, but to what extent we *are* the fieldwork. What is meant by this has been the content of this book so far, but is an incomplete, and perhaps ultimately incompletable, project. In order to know the world in a manner sufficient to live attentively as a component part of dynamic matter, it is necessary to know what surrounds us, and how complexly we interact with our world. Fieldwork, then, takes on a whole new significance as we test and interact with our sister atoms, our sibling energies, our familial beings.

But before we proceed, let us consider this intention framed from a slightly different perspective. The transverse community of action to which we belong reflects, in a wider ecosophical context, the understanding that the value of arts and humanities research is not captured by a production/consumption model, one that bears no relation to the ways in which innovation and creativity actually occur.[23] The real value of such research lies in it being:

carried by and in persons. As expertise, as confidence, as understanding and orientation to issues, problems, concerns and opportunities, as tools and abilities. This is best captured . . . in the notion of responsiveness.[24]

This responsiveness is in turn understood as an aspect of citizenship that privileges those 'spaces and opportunities for discussion, argument, critique, reflection' in which, over time, collaboration itself becomes a basis for evaluation.[25] Understanding the reasons for this emphasis on enduring collaboration and citizenship – seen as both uniting and transcending professional, disciplinary and vernacular categories – is vital to the fourth ecology as we understand it.

Such understanding ensures that, while a transverse community of action may draw heavily on skills learned in the professional worlds of the arts and the university, it remains open to, and engaged with, the full spectrum of challenges posed by the more inclusive concerns it addresses. Additionally, its members cannot simply be categorised using orthodox conventions of identity. It is neither possible nor desirable to capture these concerns in an explicit 'statement of intent' nor to paraphrase them in a core bibliography; they are by definition fluid, open and dynamically responsive to the mesh of connectivities of which they are a part. In short, their concerns correspond to the 'educational' imperatives flowing from Guattari's 'Three ecologies' and anticipated by the work of Joseph Beuys, Paulo Friere and others who work in and through the web of interconnectedness.

Concern with the relationship between creative work and a transverse community of action should not be taken as diminishing the importance of the individual, only as contesting the assumption of the sovereignty of discrete, possessively oriented individuals over their multiple attachments, connections and relationships to and with other beings, human and otherwise. Specifically, it reminds us of the misleading application of the assumptions of possessive individualism to educational and cultural work. The anthropologist James Leach reminds us that possessive individualism is the dominant secular belief system in our culture, one that requires us to take as given that individuality, creativity and originality are something exclusive to, and wholly owned by, a unique and monolithic self.[26] This belief underpins our political assumptions and social organisation through presuppositions about the natural world and human society; presuppositions reflected in our taking for granted the emphasis placed on individuals to generate novelty or originality in contemporary art and new knowledge in academic research. The position presented here takes as an alternative *a priori* that, as subjects, we are enmeshed in the world and are constituted in and through all our many and diverse forms of attachment, connectivity and relationship. It is important to note that this argument redefines the relational context and constitution of our individual becoming, but without in any sense denying validity to the individual within that meshwork.

As already indicated, Guattari's ecosophy provides one possible context in which to understand the necessity to enact, as individuals, both a greater unity and a greater differentiation regarding the self. The necessity of an *interdependent* and *porous* self that understands its own dependence on facilitating the well-being of the material, natural and social worlds, one characterised by a degree of acceptance of our multiplicity that reflects a reality grounded in diversity. That is to say, a self who is able to let go of notions of a monolithic centre with a distinct border between it and the world, that understands that it is decentralised, and that is able to pursue aims other than those of the heroic ego. The importance of any transverse community of action lies in no small part in its facilitating and sustaining such alternative and interdependent understandings of self in a culture in which artistic and academic success are modelled on the increasingly pathological and ultimately self-destructive assumptions and manifestations of a monolithically heroic ego.

Where these two versions of understanding merge is the point of departure for this book; namely, that it is imperative that to understand anything, one must return to place as a point of urgent study and action. Whether it is the fragmentation and de-centring of self, or the self-organising matter of fellow beings, or the increasing non-hierarchical structures of social dynamics that stand in conflict with economies of capitalism and growth – the imperative is one of a balanced ecosystem and sustainable existence.

At this point it is perhaps worth drawing attention to some of the characteristics of the activity of fieldwork, as regularly experienced by staff and students working in an educational context. While this will often involve the application of specific scientific and/or other methods 'in the field' – for example the use of a transect, a line set out across a natural feature along which observations are made – much of what takes place during student-focussed fieldwork is an informal building of relationships and, as such, largely unconnected to the strictly disciplinary formal agenda of the fieldwork.

In addition to their formal work, then, a group undertaking fieldwork may need, for example, to obtain provisions, establish rotas for cooking and other mundane enabling activities on which everyday life beyond the institutional basis of academy study depend. In all likelihood there will also be occasions in which a degree of mutual care is required: the treatment of blisters, cuts or insect bites, along with sharing water, tea, food and so on during rest breaks or when sheltering from adverse weather conditions. In short, the *educational* purpose of fieldwork at its best provides not simply an opportunity for students to learn some of the observational skills and hands-on practices necessary to their discipline 'out in the field'; it is also a process of socialisation through which a group – comprised of staff, students, associated experts and perhaps individuals local to the area in which the fieldwork is conducted – briefly becomes a working community embedded in the world at large (Figure 10.2).

For many students, their fieldwork trip becomes one of the most memorable aspects of their study, and for at least three reasons. First, because they learn from observing staff and others directly as both practitioners and

Figure 10.2 Rebecca Krinke, *Flood Stories* fieldwork. Photo: Rebecca Krinke.

persons, through direct and observational engagement rather than through top-down transmission. Second, because they are engaged in processes of learning a life of practical study *with* others. And third, because during fieldwork the tacit hierarchies and protocols that otherwise permeate academic relations tend to be at least partially relaxed.

In *The Power of the Ooze*, Simon Read suggests that our eco- social problems require: "a particular kind of strategy that our culture has yet to develop and promote", one that requires continuous improvisation, without a desire for perfection or a fear of failure; a situation that brings with it a particular form of joy.[27] Current Higher Education in most countries is based on a system of acquisition of set skills and content, and tested against criteria of successful proof or failure. In and of itself, the fear of failure discourages students from taking risks, from deviating from the norm, or from producing wildly innovative solutions. But we have just concluded (following Read's suggestion) that this 'deviation' from an unworkable system is exactly what is now urgently needed. One needs to reconnect with this sense of joy, of satisfaction, and the belief that in some way on is contributing to the *communitas*. If every year of a student's higher education required a project with peers from other disciplines to do one activity as collective fieldwork that benefited the environment (Figure 10.3), or, say, worked with a cultural group to change social understandings of the impact of individual behaviours, the resultant pressure on unsustainable actions across society would mandate positive changes.

'Actions speak louder than words', so goes the truism. Actions may be types of praxis, rooted in ontologies of sustainable co-existence in balance

Figure 10.3 Simon Read, *Lower Sill Intervention,* part of a stabilization project for the Falkenham Saltmarshs, Suffolk, UK. Photo: Simon Read.

with the environment. Lest this seem a blithe and naïve prognosis for the future, more easily said than done, the urgency for this action and the courage needed to undertake this radical shift should not be underestimated. As has been noted elsewhere in this book, much of this co-creative praxis in communities of shared action stands in direct opposition to the core presuppositions of large multinational companies and social institutions, to the economic interests of a consumer-led economy, and to the vested interests of many political systems. Back in 1973, E.F. Schumacher wrote *Small is Beautiful: a study of economics as if people mattered.* It offers a powerful and cogent argument against ever-increasing production and consumerism, and advocacy of 'enoughness'. Forty-five years later, we see that he was correct in predicting that ever yet more things would not increase happiness. In fact, his words stand out starkly when we learn that current World Health Organization medical statistics show that, "depression will be the second most common health problem in Western developed nations by 2020".[28] If Schumacher were alive today, he would probably be tempted to publish Volume 2, feasibly subtitled *A study of economics as if the planet mattered.* With a combination of influences from Mahatma Ghandi, to Buddhism, Karl Marx, Catholicism, and studying under his mentor Leopold Kohr, Schumaker espoused 'appropriate technology' and humanising the work force to restore the dignity of work. And finally, he advocated the nationalised (Socialist) running of energy companies, with the absolute ability to control and out-capitalise the capitalists . . . to evolve a more democratic and dignified system of industrial administration, a more humane employment of machinery, and a more intelligent utilisation of the fruits of human ingenuity and effort.[29]

Again, we risk despair if we focus too exclusively on the mistakes of consumerism, and our failure to learn lessons from the past. Instead, we will firmly turnabout and resolutely find the way forward, through positive actions, through attentiveness and living lightly on the earth, shrinking carbon footprints and restoring depleted habitats. To return to the introduction to this chapter, this is the *ethos* we wish to re-establish, how we wish to raise our children and how we wish to leave our communities better places than we found them.

To live is to become, and to become requires a place in which to exist.[30] All life is quintessentially sited life; there is no such thing as life lived in the abstract. Acknowledging the need to be somewhere, it follows that the 'stuff' of this somewhere must be the air we breathe, the ground upon which we stand, the shelter that protects us, the clothing that keeps us warm, the water we drink and the food we eat to sustain life. Through reimagined fieldwork, and continual efforts to preserve and protect these basic elements, the living balance of our relationship with all else on this earth can be addressed and our joyful living restored. Reimagining can also help us reconfigure priorities, starting by asserting the values collaborative communities who work together, by collectively honouring the sites of *communitas* with human and non-human beings, and sharing through the recognition of mutual needs.

This reimagining will occur when an educational system builds on associative and creative learning, understanding that lateral thinking and active practices make for an enthusiastic learner, one who sees connections and is inspired to try new things without a fear of failure. Understanding that diversity and multiplicity are assets through which many perspectives and skills are combined as a strength will enable those who feel different to also feel that they too have a place in the new ethos of ecosophy. There is wisdom to be found in the various Indigenous communities across the world. Peoples who have lived in close harmony with their environment over many generations have usually arrived at a point of deep knowledge, shared with stories and practices that have had iterations over the centuries, in patterns that sustain a balance among humans and non-humans alike. A commitment to re-indigenising knowledge is one that seeks to respect this knowledge and derive practical understandings from it. In this we share with Arturo Escobar and many others the understanding of the importance of movements such as the Zapatista, with their 'concept of the pluriverse, *a world where many worlds fit*'.[31] In short, a new conception of fieldwork that seeks to re-indigenise knowledge, while it will learn valuable lessons from those who articulate a "decolonial political vision of a world in which many worlds would coexist",[32] will nonetheless need to remain firmly grounded in understandings of the particularities of the local world in which it is exercised.[32] It requires, that is, a politics of contingency, a 'creative politics of "bricolage" . . . grounded in the local and historical specificities of particular people, places and times'; that can exercise the testimonial imagination necessary to keep alive "local memory and imagination as a reservoir of meanings, truths, and possibilities for a different future".[33] At least one example of the rethinking of fieldwork in practice we are arguing for here has already been set out by the artist and philosopher turned anthropologist, A. David Napier, long and insightful critic of the failings latent in so much contemporary art.[34] Those who wish to explore what we have outlined here in more detail are, in consequence, strongly advised to read and take up the exercises set out in Napier's *Making Things Better* as a basis for developing a fieldwork appropriate to their own community, place and time.[35] With this thought we end this chapter and this book, hopeful that these words will ring true now and in times to come and, equally hopeful as well that shared communities of transverse creative practices will continue to grow and affect positive change.

Reimagined fieldwork is, then, the manner by which we turn from absorbing the hard lessons of the past to a more hopeful, more collaborative and active praxis that can first imagine, and then create, a more sustainable future.

Notes

1 Edward S. Casey, *The Fate of Place: A Philosophical History* (Berkeley: University of California Press, 1997), xiv.
2 Casey, *The Fate of Place*, citing Victor Turner, *The Ritual Process: Structure and Anti-Structure* (Chicago, IL: Aldine, 1969). Casey opposes Turner's position with that of Jean Luc-Nancy, who states: '*In place of* community there is now no *place*, no site, no temple or altar for community. Exposure takes place everywhere, in all places, for it is the exposure of all and each in his solitude, to not being alone'.'. [Jean-Luc Nancy, *The Inoperative Community*, trans. by Peter Connor and others (Minneapolis: University of Minnesota Press, 1991), 143, his italics.]
3 Amitav Ghosh, *The Great Derangement: Climate Change and the Unthinkable* (Chicago, IL: University of Chicago Press, 2017).
4 Bruno Latour, *Down to Earth: Politics in the New Climatic Regime*, trans. by Catherine Porter (Cambridge, MA: Polity Press, 2018).
5 Jo Gullies, 'The Spatial Turn', <http://www.spatial.scholarslab.org>.
6 Gullies, 'The Spatial Turn', on behalf of the Institute for the Enabling of Geopolitical Scholarship (2011).
7 Donna Haraway, *Staying with the Trouble*: *Making Kin in the Cthulhucene* (Durham, NC: Duke University Press, 2016).
8 Anselm Franke, 'Introduction—"Animism"', *e-flux journal*, 36 (July 2012), 2.
9 Mary Watkins, '"Breaking the Vessels": Archetypal Psychology and the Restoration of Culture, Community and Ecology', in *Archetypal Psychologies: Reflections in Honor of James Hillman*, ed. by Stanton Marlan (New Orleans, LA: Spring Journal Books, 2008), 425.

10 James Hillman, 'Man Is By Nature a Political Animal, or Patient as Citizen', in *Speculations after Freud: Psychoanalysis, philosophy and culture*, eds. Sonu Shamdasani and Michael Munchow (London: Routledge, 1994), 33.

11 Mary Watkins and Helene Shulman, *Towards Psychologies of Liberation* (New York: Palgrave Macmillan, 2008), 10.

12 Edward Sampson, as quoted in Hillman 'Man Is By Nature a Political Animal', 32.

13 Bernd Reiter, ed., *Constructing the Pluriverse: The Geopolitics of Knowledge* (Durham, NC: Duke University Press, 2018), 9.

14 Geraldine Finn, *Why Althusser Killed His Wife: Essays on Discourse and Violence* (Atlantic Highlands, NJ: Humanities Press, 1996), 145.

15 Michael Dear and Others, *GeoHumanities: Art, History, Text at the Edge of Place* (Abingdon: Routledge, 2011), 9.

16 This is almost a verbatim quote from Ralph Ellison who wrote in *Invisible Man* 'if you don't know where you are, you probably don't know who you are'.'. [Ralph Elision, *Invisible Man* (New York: Random House, 1952), 564.] This is also close to the words of Wendell Berry, who says: "If you don't know where you are you don't know who you are", as quoted in Wallace Stegner, *The Sense of Place: A Nine Page Pamphlet* (Madison: Wisconsin Humanities Committee, 1986), 1.

17 Walter Benjamin and Asja Lācis, "Neapel ('Naples')", *Frankfurter Zeitung* (19 August 1925), 166. The emphasis here, as they discuss the city of Naples, is less on barriers or dividing lines between neighbourhoods, and more on qualities of recognition, especially in areas with multiple characteristics.

18 Mary Modeen, "Breaking the Boundaries of 'Self': Representations of Spatial Indeterminacy", *Architecture and Culture Journal*, special issue *Transgression: Body and Space*, 2, no. 3 (November 2014), 332–358.

19 Victor Burgin, *In/different Spaces: Place and Memory in Visual Culture* (Berkeley: University of California Press, 1996), 115.

20 Graeme Gilloch, *Myth and Metropolis: Walter Benjamin and the City* (Cambridge, MA: Polity Press, 1996), quoted by Jeff Matthews: 'Walter Benjamin and Naples', *Naples: Life, Death and Miracles* (29 March 2016) <http://www.naplesldm.com/benjamin.php> [accessed 10 April 2020].

21 Luce Irigaray, *An Ethics of Sexual Difference*, trans. by Carolyn Burke and Gillian Gill (Ithaca, NY: Cornell University Press, 1982). Irigaray writes:

> We must, therefore, reconsider the whole of our conception of place, both in order to move onto another age of difference (each age of thought corresponds to a particular time of gender and difference), and in order to construct an ethics of passions . . .
>
> (*An Ethics of Sexual Difference*, 12–13)

22 Ken Kiff, letter to Iain Biggs, 29th May 1998, as quoted in Emma Hill, *Ken Kiff: The Sequence* (Norwich: Sainsbury Centre for Visual Arts, 2018), 24.

23 James Leach and Lee Watson, 'Enabling Innovation: Creative Investments in Arts and Humanities Research' (2010), <http://www.jamesleach.net/articles.html 2010> [accessed 30 March 2013].

24 Leach and Watson, 'Enabling Innovation', 6.

25 Ibid., 3.

26 James Leach, 'Creativity, Subjectivity and the Dynamic of Possessive Individualism', in *Creativity and Cultural Improvisation*, eds. Elizabeth Hallam and Tim Ingold (Oxford & New York: Berg, 2007), 99–116.

27 Simon Read, 'The Power of the Ooze', in *The Power of the Sea: Making Waves in British Art, 1790–2014*, ed. by Janette Kerr and Christina Payne (Bristol: Sansom & Co Ltd. 2014), 46.

28 Madeleine Bunting, 'Small Is Beautiful: An Economic Idea That Has Sadly Been Forgotten', *The Guardian* (10 November 2011).

29 E. F. Schumacher, *Small Is Beautiful: A Study of Economics as if People Mattered* (London: Blond and Briggs, 1975), 261.

30 This simple formulation underlays the theory of *dasein*, literally 'living in place', discussed earlier in this book.

31 Arturo Escobar, *Designs for the Pluriverse: Radical Interdependence, Autonomy and the Making of Worlds* (Durham, NC: Duke University Press, 2018), xvi.

32 Reiter, *Constructing the Pluriverse*, ix.

33 Finn, *Why Althusser Killed His Wife*, 145.

34 See, for example, the first two chapters of A. David Napier, *Foreign Bodies: Performance, Art and Symbolic Anthropology* (Oakland: University of California Press, 1992).

35 David Napier, *Making Things Better: A Workbook on Ritual, Cultural Values, and Environmental Behavior* (New York: Oxford University Press, 2014).

Bibliography

Allen, Richard J., Graham J. Hitch, and Alan D. Baddeley, 'Cross-Modal Binding and Working Memory', *Visual Cognition*, 17, no. 1–2 (2009), 83–102, <http://www.tandfonline.com/li/abs/10.1080/13506280802281386>.

Ault, Alicia, 'Ask Smithsonian: What's the Deepest Hole Ever Dug?', *Smithsonianmag* (2015), <https://www.smithsonianmag.com/smithsonian-institution/ask-smithsonian-whats- deepest-hole-ever-dug-180954349/> [accessed 10 April 2020].

Australian Government, 'Indigenous Protected Areas', *Department of Agriculture, Water and the Environment*, <http://www.environment.gov.au/land/indigenous-protected-areas> [accessed 10 April 2020].

——, 'Lake Condah', Department of Agriculture, Water and Environment, <https://www.environment.gov.au/indigenous/ipa/declared/lake-condah.html> [accessed 16 November 2018].

Bachelard, Gaston, *The Poetics of Space*, trans. by Maria Jolas (Boston, MA: Beacon Press, 1958; repr. 1994).

Bailey, Jane, and Iain Biggs, '"Either Side of Delphy Bridge": A Deep Mapping Project Evoking and Engaging the Lives of Older Adults in Rural North Cornwall', *Journal of Rural Studies*, 28, no. 4 (2012), 318–328, <http://dx.doi.org/10.1016/j.jrurstud.2012.01.001>.

Bailey, Jane, Iain Biggs, and Dan Buzzo, 'Deep Mapping and Rural Connectivities', in *Grey and Pleasant Land? Older People's Connectivity in Rural Community Life*, ed. by Catherine Hagan Hennessey, Robin Means, and Vanessa Burholt (Bristol: Policy Press, 2014), 159–192.

Barad, Karen, *Meeting the Universe Halfway: Quantum Physics and the Entanglement of Matter and Meaning* (Durham, NC: Duke University Press, 2007).

Barnes, Luci Gorell, 'The Atlas of Human Kindness', <http://www.lucigorellbarnes.co.uk/the-atlas-of-human-kindness/> [accessed 10 April 2020].

——, 'Luci Gorell Barnes', <http://www.lucigorellbarnes.co.uk> [accessed 10 April 2020].

——, 'This Long River', in *Water, Creativity and Meaning: Multidisciplinary Understandings of Human-Water Relationships*, ed. by Liz Roberts and Katherine Phillips (London: Routledge, 2018), 36–53, Taylor & Francis e-book.

Bauman, Zygmunt, and Leonidas Donskis, *Moral Blindness: The Loss of Sensitivity in Liquid Modernity* (Cambridge, MA: Polity Press, 2013).

Bdote Memory Map, 'Bdote Memory Map: Mitakuyepi! Welcome!', <http://bdotememorymap.org/> [access 10 April 2020].

Benjamin, Walter, 'Little History of Photography', in *Walter Benjamin: Selected Writings: Vol. 2: 1927–1934*, ed. by Michael W. Jennings (Cambridge, MA: The Belknap Press of Harvard University, 1999), 507–530.

Benjamin, Walter, and Asja Lacis, 'Naples', in *Reflections: Essays, Aphorisms, Auto-biographical Writings*, trans. by Edmund Jephcott, ed. by Peter Demetz (New York: Schocken, 1978), 166–167.

Bennett, Jane, *Vibrant Matter: A Political Ecology of Things* (Durham, NC: Duke University Press, 2010).

Biggs, Iain, *Between Carterhaugh and Tamshiel Rig: A Borderline Episode* (Bristol: Wild Conversations Press for TRACE, 2004).

———, '"Incorrigibly Plural?" Rural Lifeworlds between Concept and Experience', *Canadian Journal of Irish Studies*, 38, no. 1–2 (2014), 260–279.

———, 'The Spaces of "Deep Mapping": A Partial Account', *Journal of Arts and Communities*, 2, no. 1 (2011), 5–25.

Bissell, Laura, and David Overend, 'Regular Routes: Deep Mapping a Performative Counterpractice for the Daily Commute', in *Deep Mapping*, ed. by Les Roberts (Basel: MDPI AG – Multidisciplinary Digital Publishing Institute, 2016), 131–154.

Blainey, Geoffrey, *The Story of Australia's People: The Rise and Fall of Ancient Australia* (Melbourne: Penguin, 2015).

Bly, Robert, 'Surprised by Evening', in *Silence in the Snowy Fields: Poems* (Middletown, CT: Wesleyan University Press, 1962), 15.

Bodhi, Bhikkhu, *The Noble Eightfold Path: The Way to the End of Suffering* (Kandy: Buddhist Publication Society, 1994).

The Booke of the Universal Kirk of Scotland: Acts and Proceedings of the General Assemblies of the Kirk of Scotland from the Year MDLX, ed. by Church of Scotland General Assembly (Edinburgh, 1845).

Borthwick, David, Pippa Marland, and Anna Stenning, eds, *Walking, Landscape and Environment* (London: Routledge, 2020).

Bové, Paul, *Mastering Discourse: The Politics of Intellectual Culture* (Durham, NC: Duke University Press, 1992).

British National Party, 'Stopping All Immigration', <https://bnp.org.uk/policies/immigration/> [accessed 10 April 2020].

Bruno, Giuliana, *Atlas of Emotion: Journeys in Art, Architecture, and Film* (London: Verso, 2002).

Bryson, Bill, *A Walk in the Woods: Rediscovering America on the Appalachian Trail* (New York: Broadway Books, 1997).

Bunting, Madeleine, 'Small Is Beautiful: An Economic Idea That Has Sadly Been Forgotten', *Guardian* (10 November 2011), n.p.

Burgin, Victor, *In/different Spaces: Place and Memory in Visual Culture* (Berkeley: University of California Press, 1996).

Burnett, Kathryn A., 'Place Apart: Scotland's North as a Cultural Industry of Margins', in *Relate North: Culture, Community and Communication*, ed. by Timo Jokela and Glen Coutts (Rovaniemi: Lapland University Press, 2017), 60–83.

Butler, Judith, *Antigone's Claim: Kinship Between Life and Death* (New York: Columbia University Press, 2000).

Cardiff, Janet, and George Bures Miller, *Lost in the Memory Palace: Janet Cardiff and George Bures Miller*, 6 April–18 August 2013, exhibition at The Art Gallery of Ontario, <http://ago.ca/exhibitions/lost-memory-palace-janet-cardiff-and-george-bures- miller> [accessed 10 April 2020].

Carson, Rachel, *Silent Spring* (Boston, MA: Houghton Mifflin, 1962).

Casey, Edward S., *The Fate of Place: A Philosophical History* (Berkeley: University of California Press, 1997).

———, *Getting Back into Place: Toward a Renewed Understanding of the Place-World* (Bloomington: Indiana University Press, 1993).

Causey, Andrew, *Peter Lanyon: Modernism and the Land* (London: Reaktion Books, 2006).

Center for Art and Social Engagement, 'Centering Creativity, Impact and Community', *University of Houston*, <https://uh.edu/kgmca//case/> [accessed 10 April 2020].

Chai, Paul, 'On a Mission: Uncovering the Past of Victoria's Gunditjmara Country', *Traveller* (2017), <http://www.traveller.com.au/on-a-mission-uncovering-the-past-of-victorias- gunditjmara-country-gtvi9t> [accessed 10 April 2020].

Chetwynd, Courtney, '"It Tells Us": The Practice of Performativity and Liminality within Northern Culture' (unpublished doctoral dissertation, University of Dundee, 2018).

'Chorography', *Merriam-Webster.com Dictionary*, <https://www.merriam-webster.com/dictionary/chorography> [accessed 10 April 2020].

Conradie, Hanien, *The Voice of Water: Re-Sounding a Silenced River*, Presented at the *Liquidscapes* conference, Dartington, England (2018).

Coole, Diana, and Samantha Frost, eds, *New Materialisms: Ontology, Agency and Politics* (Durham, NC: Duke University Press, 2010).

Courage, Cara, 'Dr Cara Courage: Head of Tate Exchange at Tate', *LinkedIn*, <https://www.linkedin.com/in/caracourage/> [accessed 10 April 2020].

———, *Placemaking and Community*, Online Video Recording, YouTube (TEDx Indianapolis, 2017), <https://www.youtube.com/watch?v=Sfk1ZW9NRDY> [accessed 10 April 2020].

Coverley, Merlin, *Psychogeography* (Harpenden: Oldcastle Books, 2006).

Cresswell, Tim, *Place: A Short Introduction* (Oxford: Blackwell Publishing, 2004).

Crist, Meehan, 'Besides I'll Be Dead', *London Review of Books*, 40, no. 4 (22 February, 2018), 12.

Cronin, Michael, 'Who Fears to Speak in the New Europe? Plurilingualism and Alterity', *European Journal of Cultural Studies*, 15, no. 2 (2012), 242–257.

Deakin, Roger, *Wildwood: A Journey through Trees* (London: Hamish Hamilton, 2007).

Dear, Michael, et al., eds. *GeoHumanities: Art, History, Text at the Edge of Place* (Abingdon: Routledge, 2011).

de Certeau, Michel, *The Practice of Everyday Life* (London: University of California Press, 1988).

Demos, T. J., *Decolonizing Nature: Contemporary Art and the Politics of Ecology* (Berlin: Sternberg Press, 2016).

Derrida, Jacques, *Writing and Difference*, trans. by Alan Bass (Chicago, IL: University of Chicago Press, 1978).

de Ville, Nicholas, and Stephan Foster, eds, *The Artist and the Academy: Issues in Fine Art Education and the Wider Cultural Context* (Southampton: John Hansard Gallery, 1994).

Dewey, John, *Experience and Education* (New York: Collier, 1938).

Dewsbury, John-David, 'Witnessing Space: "Knowledge without Contemplation"', *Environment and Planning*, 35 (2003), 1907–1932.

Dolphijn, Rick, and Iris Van Der Tuin, *New Materialism: Interviews and Cartographies* (Ann Arbor: University of Michigan Press, 2012).

Drengson, Alan, 'Some Thought on the Deep Ecology Movement', *Foundation for Deep Ecology*, <http://www.deepecology.org/deepecology.htm> [accessed 10 April 2020].

Drengson, Alan, and Bill Devall, eds, *The Ecology of Wisdom: Writings by Arne Naess* (Berkeley, CA: Counterpoint, 2008).

Dreyfus, Hubert, *Being-in-the-World* (Cambridge: MIT Press, 1995).

Drucker, Johanna, *Sweet Dreams: Contemporary Art and Complicity* (Chicago, IL: University of Chicago Press, 2005).

Dubos, Rene, *So Human an Animal: How We Are Shaped by Surroundings and Events* (New York: Scribner, 1968).

Ehrlich, Paul R., and Anne H. Ehrlich, *The Dominant Animal: Human Evolution and the Environment* (Washington, DC: Island Press, 2009).

Elision, Ralph, *Invisible Man* (New York: Random House, 1952).

Elkins, James, 'Writing Moods', in *Landscape Theory*, ed. by Rachel Ziady DeLue and James Elkins (London: Routledge, 2008). 69–86.

English, Darby, 'Modernism's War on Terror', in *Outliers and American Vanguard Art*, ed. by Lynne Cooke (Chicago, IL: University of Chicago Press, 2018). 31–41.

Escobar, Arturo, *Designs for the Pluriverse: Radical Interdependence, Autonomy and the Making of Worlds* (Durham, NC: Duke University Press, 2018).

Evans, Brad, and Julian Reid, *Resilient Life: The Art of Living Dangerously* (Hoboken, NJ: Wiley, 2014).

Favell, Adrian, 'Socially Engaged Art in Japan: Mapping the Pioneers', *Field*, 7 (2017), <http://field-journal.com/issue-7/socially-engaged-art-in-japan-mapping-the-pioneers> [accessed 10 April 2020].

Finn, Geraldine, *Why Althusser Killed His Wife: Essays on Discourse and Violence* (Atlantic Highlands, NJ: Humanities Press, 1996).

Fiumara, Gemma Corrodi, *The Other Side of Language: A Philosophy of Listening*, trans. by Charles Lambert (London: Routledge, 1995).

———, *Psychoanalysis and Creativity in Everyday Life: Ordinary Genius*, trans. by Charles Lambert (London: Routledge, 2013).

Fowle, Kate, 'Introduction', in Terry E. Smith, *Thinking Contemporary Curating* (Berkeley, CA: University of California, Independent Curators International, 2012), 7–15.

Frampton, Kenneth, 'Place-Form and Cultural Identity', in *Design after Modernism*, ed. by John Thackara (London: Thames & Hudson, 1988), 51–66.

Franke, Anselm, 'Introduction—"Animism"', *e-flux journal*, 36 (July 2012), 1–2.

Friere, Paolo, *Pedagogy of the Oppressed*, trans. by Myra Raomos (New York: Bloomsbury, 1970).

Frodeman, Robert, *Geo-Logic: Breaking Ground between Philosophy and the Earth Sciences* (Albany: State University of New York Press, 2003).

Fung, Stanislaus, 'Four Key Terms in the History of Chinese Gardens', trans. by Mark Jackson, Paper presented at the International Conference on Chinese Architectural History, 7–10 August 1995.

Gablik, Suzy, 'Connective Aesthetics Art after Individualism', in *Mapping the Terrain: New Genre Public Art*, ed. by Suzanne Lacy (Seattle, WA: Bay Press, 1995), 74–83.

Gare, Arran. E., *Postmodernism and the Environmental Crisis* (London: Routledge, 1995).

Garner, Alan, *The Voice That Thunders* (London: The Harvill Press, 1997).

Gellner, Ernest, *Postmodernism, Reason and Religion* (London: Routledge, 1992).

Genosko, Gary, 'A-Signifying Semiotics', *The Public Journal of Semiotics*, 2, no. 1 (2008), 11–21.

Ghosh, Amitav, *The Great Derangement: Climate Change and the Unthinkable* (Chicago, IL: University of Chicago Press, 2017).

Gibson, David, 'Joseph Beuys 40th Anniversary Journey "to The Moor of Rannoch and the Road to the Isles"', *Studio International* (8 May 2010), <http://www.studiointernational.com/index.php/joseph-beuys-40th-anniversary-journey-to-the-moor-of-rannoch-and-the-road-to-the-isles> [accessed 10 April 2020].

Gilloch, Graeme, *Myth and Metropolis*: *Walter Benjamin and the City* (Cambridge, MA: Polity Press, 1996).

Glück, Louise, *Firstborn* (New York: Harper Collins, 1968) and *The First Five Books of Poems* (Manchester: Carcanet Press, 2014).

Goodell, Jeff, *The Waters Will Come: Rising Seas, Sinking Cities and the Remaking of the Civilised World* (New York: Little, Brown and Company, 2017).

Grant, E.K., *Unseen, Unheard, Unspoken: Exploring the Relationship between Aboriginal Spirituality and Community Development* (Adelaide: University of South Australia, 2004), 8–9.

Gregory, Ian, et al., 'Spatializing and Analyzing Digital Texts: Corpora, GIS and Places', in *Deep Maps and Spatial Narratives*, ed. by David Bodenhamer, John Corrigan, and Trevor Harris (Bloomington: Indiana University Press, 2015), 150–178.

Gros, Frederic, *A Philosophy of Walking*, trans. by John Howe (London: Verso, 2015).

Groth, Paul, 'Frameworks for Cultural Landscape Study', in *Understanding Ordinary Landscapes*, ed. by Paul Groth and Todd W. Bressi (New Haven, CT: Yale University Press, 1997), pp. 1–24.

Guattari, Felix, *The Three Ecologies*, trans. and ed. by Ian Pindar and Paul Sutton (London: Bloomsbury, 2005).

Guldi, Jo, 'What Is the Spatial Turn?', *University of Virginia Library: Scholar's Lab*, <http://spatial.scholarslab.org/> [accessed 10 April 2020].

Haberle, Simon G., and Bruno David, eds, 'Peopled Landscapes: Archaeological and Biogeographic Approaches to Landscapes', *Terra Australis*, 34 (Canberra: ANU E Press, 2012), 103–120.

Hanrahan, Catherine, 'Chart of the Day: Is this the Most Severe Drought in History?', *ABC News On-line* (6 August 2018), <https://www.abc.net.au/news/2018-08-06/how-does-the-current-drought-compare/10055414> [accessed 10 April 2020].

Haraway, Donna J., *Staying with the Trouble: Making Kin in the Chthulucene* (Durham, NC: Duke University Press, 2016).

Harding, Stephen, 'What Is Deep Ecology?', *Schumacher College*, <https://www.schumachercollege.org.uk/learning-resources/what-is-deep-ecology> [accessed 10 April 2020].

Harmon, Katherine, *You Are Here: Personal Geographies and Other Maps of the Imagination* (New York: Princeton Architectural Press, 2004).

Harrison, Robert Pogue, 'Hic Jacet', in *Landscapes and Power*, ed. by W. J. T. Mitchell (Chicago, IL: University of Chicago Press, 2002), 349–364.

Healy, Ciara, and Adam Stead, *Hold, Test, Empty, Remove, Repeat*, 2018, video, 21 min., <https://vimeo.com/213206929> [accessed 10 April 2020].

Heat-Moon, William Least, *PrairyErth: A Deep Map* (Boston, MA: Houghton Mifflin, 1991).

Heddon, Dee, and Misha Myers, 'The Walking Library for Women', in *Walking, Landscape and Environment*, ed. by David Borthwick, Pippa Marland, and Anna Stenning (London: Routledge, 2020). 113–126.

Heelas, Paul, and Linda Woodhead, *The Spiritual Revolution: Why Religion is Giving Away to Spirituality* (Hoboken, NJ: John Wiley & Sons, 2004).

Heidegger, Martin, *Being and Time* (Oxford: Blackwell, 1962).

———, *Poetry, Language, Thought*, trans. by Albert Hofstadter (London: Harper Collins, 2001).

———, 'The Thing', in *What Is a Thing?*, trans. by William Baynard Barton and Vera Deutsch, 6137 (Washington, DC: Henry Regnery Co., 1969).

Hess, Harry, 'History of Ocean Basins', in *Petrologic Studies: A Volume in Honor of A. F. Buddington*, ed. by A. E. J. Engel, Harold L. James, and B. F. Leonard (Boulder, CO: Geological Society of America, 1962), 599–620.

Hill, Emma, *Ken Kiff: The Sequence* (Norwich: Sainsbury Centre for Visual Arts, 2018).

Hiller, Susan, *The Last Silent Movie*, 2007–08, Audio-Visual Work, 22-mins., British Council Collection.

Hillman, James, 'Man Is by Nature a Political Animal, or Patient as a Citizen', in *Speculations after Freud: Psychoanalysis, Philosophy and Culture*, ed. by Sonu Shamdasani and Michael Munchow (London: Routledge, 1994), 27–40.

———, *The Thought of the Heart* (Dallas, TX: Spring Publications, 1984).

Holub, Miroslav, *The Dimensions of the Present and Other Essays*, ed. by David Young (London: Faber & Faber, 1990).

Huizinga, Johan, *Homo Ludens* (Boston, MA: Beacon Hill Press, 1955).

Hurd, Barbara, *Stirring the Mud: On Swamps, Bogs and Human Imagination* (Athens: University of Georgia Press, 2008).

Husserl, Edmund, and Rochus Sowa, *Die Lebenswelt* (New York: Springer, 2008).

Illich, Ivan, *Disabling Professions* (London: Marion Boyers, 1977).

Ingold, Tim, *Being Alive: Essays on Movement, Knowledge and Description* (London: Routledge, 2011).

———, *The Perception of the Environment: Essays in Livelihood, Dwelling and Skill* (London: Routledge, 2000).

Irigaray, Luce, *An Ethics of Sexual Difference*, trans. by Carolyn Burke and Gillian Gill (Ithaca, NY: Cornell University Press, 1982).

———, *To Speak Is Never Neutral* (London: Routledge, 2002).

Jacob, Mary Jane, McKnight Visual Artists Fellowship Exhibition catalogue (Minneapolis, MN: MCAD, 2012). <https://www.artandeducation.net/announcements/109281/mcknight-visual-artists-fellowship-exhibition>

Jamie Forbert Architects, 'Patrick Keiller: The Robinson Institute', <https://jamiefobertarchitects.com/work/patrick-keiller/> [accessed 10 April 2020].

Jones, Anwen, and Rowan O'Neill, 'Living Maps of Wales: Cartography as Inclusive, Cultural Practice in the Works of Owen Rhoscomyl (Arthur Owen Vaughan) and Cliff McLucas', *International Journal of Welsh Writing in English*, 2, no. 1 (October 2014), 106–123.

Jones, Owain, 'Stepping from the Wreckage: Geography, Pragmatism and Anti-representational Theory', *Geoforum*, 39 (July 2008), 1600–1612.

———, 'True Geography [ ] Quickly Forgotten, Giving Away to an Adult-Imagined Universe. Approaching the Otherness of Childhood', *Children's Geographies: Emerging Issues in Children's Geographies Lorraine Van Blerk and John Barker*, 6, no. 2 (2008), 195–212. <https://www.tandfonline.com/doi/abs/10.1080/14733280801963193>

Jones, Owain, and Paul Cloke, *Tree Cultures: The Place of Trees and Trees in Their Place* (London: Bloomsbury Academic, 2002).

Jonker, Julian, and Karen E. Till, 'Mapping and Excavating Spectral Traces in Post-Apartheid Cape Town', *Memory Studies*, 2, no. 3 (2009) 303–335.

Jung, Carl, *The Collected Writings of C.G. Jung*, 20 vols, 2nd edn (Princeton, NJ: Princeton University Press, 1970).

———, *The Earth Has a Soul: C.G. Jung on Nature, Technology and Modern Life*, ed. by Meredith Sabini (Berkeley, CA: North Atlantic Books, 2016).

Ka'ai, Tania, and Others, *Ki Te Whaio: An Introduction to Māori Culture and Society* (Auckland: Pearson, 2004).

Kavanagh, Erin, 'Re-thinking the Conversation: A Geomythological Deep Map', in *Re-Mapping Archaeology: Critical Perspectives, Alternative Mappings*, ed. by Mark Gillings, Piraye Hacıgüzeller and Gary Lock (London: Routledge, 2019), 201–230.

Kaye, Nick, *Site-Specific Art: Performance, Place and Documentation* (Milton Park: Routledge, 2000).

Keiller, Patrick, *The Robinson Institute*, 27 March–14 October 2012, Installation Exhibition, Tate Britain, <https://www.tate.org.uk/whats-on/tate-britain/exhibition/patrick-keiller-robinson-institute> [accessed 10 April 2020].

Kraynak, Janet, 'The Land and the Economics of Sustainability', *Art Journal*, 69, no. 4 (2010), 16–25.

Kwon, Miwon, *One Place after Another: Site Specific Art and Locational Identity* (Cambridge: The MIT Press, 2004).

Lakoff, George, 'Why It Matters How We Frame the Environment', *Environmental Communication*, 4, no. 1 (2010), 70–81.

The Land Foundation, 'The Land Foundation', <https://www.thelandfoundation.org/> [accessed 10 April 2020].

Latour, Bruno, *Down to Earth: Politics in the New Climatic Regime*, trans. by Catherine Porter (Cambridge, MA: Polity Press, 2018).

———, *An Inquiry into Modes of Existence*, trans. by Catherine Porter (Cambridge, MA: Harvard University Press, 2013).

Lavery, Cari, 'Walking and Theatricality: An Experiment', in *Walking, Landscape and Environment*, ed. by David Borthwick, Pippa Marland, and Anna Stenning (London: Routledge, 2020), 36–50.

Leach, James, 'Creativity, Subjectivity and the Dynamic of Possessive Individualism', in *Creativity and Cultural Improvisation*, ed. by Elizabeth Hallam and Tim Ingold (Oxford: Berg, 2007), 99–116.

Leach, James, and Lee Watson, *Enabling Innovation: Creative Investments in Arts and Humanities Research* (London: Arts and Humanities Research Council, 2010).

Lee, Gini, 'The Intention to Notice: The Collection, the Tour and Ordinary Landscapes' (unpublished doctoral dissertation, RMIT University, 2006).

Lefebvre, Henri, *The Production of Space*, trans. by Donald Nicholson-Smith (Oxford: Blackwell, 1991).

Le Guin, Ursula K., 'Freedom', in *Words Are My Matter: Writings about Life and Books, 2000–2016* (Easthampton, MA: Small Beer Press, 2016).

Le Guin, Ursula K., Todd Barton, and Margaret Chodos-Irvine, *Always Coming Home* (Berkeley: University of California Press, 1986).

Lippard, Lucy, *The Lure of the Local: Senses of Place in a Multicentered Society* (New York: New Press, 1997).

———, *Overlay: Contemporary Art and the Art of Prehistory* (New York: Pantheon Books, 1983).

———, 'Place and Histories: Writing Other People's Memories', in *The Intelligence of Place: Topographies and Poetics*, ed. by Jeff Malpas (London: Bloomsbury, 2015), 51–64.

Loeffler, Sylvia, 'Glas Journal 2015', <https://silvialoeffler.wordpress.com/about/transit-gateway-4-frenzy-and-excitement/transit-gateway-3-structures-of-care/transit-gateway-2-walls-of-protection/transit-gateway-1-a-shoreline-of-anxiety/glas-journal-2016/glas-journal-2015/> [accessed 10 April 2020].

———, '*Glas Journal*: Deep Mappings of a Harbour or the Charting of Fragments, Traces and Possibilities', in *Deep Mapping*, ed. by Les Roberts (Basel: MDPI AG – Multidisciplinary Digital Publishing Institute, 2016), 30–48.

———, 'Place Values – *Glas Journal*: A Deep Mapping of Dún Laoghaire Harbour (2014–2016)', in *Landscape Values: Place and Praxis*, ed. by Tim Collins, and others (Galway: Centre for Landscape Studies NUI Galway, 2016), 170–173.

Long, Richard, *A Line Made by Walking*, Documentation at the Tate Britain, London, 1967.

Lorimer, Hayden, 'Walking: New Forms and Spaces for Studies of Pedestrianism', in *Geographies of Mobilities: Practices, Spaces and Subjects*, ed. by Tim Cresswell and Peter Merriman (London: Routledge, 2010), 19–34.

Luchte, James, *Of the Feral Children* (London: Createspace, 2012).

MacFarlane, Robert, *The Old Ways: A Journey on Foot* (London: Hamish Hamilton, 2012).

Majozo, Estella Conwill, 'To Search for the Truth and Make It Matter', in *Mapping the Terrain: New Genre Public Art*, ed. by Suzanne Lacy (Seattle, WA: Bay Press, 1995), 88–93.

Malafouris, Lambros, *How Things Shape the Mind: A Theory of Material Engagement* (Cambridge: The MIT Press, 2013).

Maris, Alexander, and Susan Maris, *Uriel*, 2008, Single Channel Video, 34 mins, 8 secs, <https://www.fvu.co.uk/projects/uriel> [accessed 10 April 2020].

Massey, Doreen, *For Space* (London: SAGE Publications, 2005).

———, 'A Global Sense of Place', *Marxism Today*, 38 (1991), 24–29.

———, 'Landscape as a Provocation: Reflections on Moving Mountains', *Journal of Material Culture*, 11, no. 33 (2006), 33–48.

Matthews, Jeff, 'Walter Benjamin and Naples', *Naples: Life, Death and Miracles* (29 March 2016), <http://www.naplesldm.com/benjamin.php> [accessed 10 April 2020].

McGrath, Ann, and Mary Anne Jebbs, eds, *Long History, Deep Time: Deepening Histories of Place* (Canberra: Australian National University Press, 2015).

McKay, Don, 'Astonished', in *Strike/Slip* (Toronto: McLelland and Stewart, 2006), 3.

———, 'Some Remarks on Poetry and Poetic Attention', in *The Second MacMillan Anthology*, ed. by John Metcalf and Leon Rooke (Toronto: MacMillan, 1989), 206–208.

———, *Vis-à-vis: Fieldnotes on Poetry and Wilderness* (Wolfville: Gaspereau Press, 2001).

McLucas, Cliff, 'Deep Mapping', *CliffordMcLucas*, <http://cliffordmclucas.info/deep-mapping.html> [accessed 10 April 2020].

———, 'I Was Invited to This Island', in *Cliff McLucas Collection*, Audio Recording, National Library of Wales collection [n.d.]. MCLT, National Library of Wales, Aberystwyth <http://www.archaeographer.com/keyword/cliff%20mclucas/> [accessed 10 April 2020].

McManus, Tony, *The Radical Field: Kenneth White and Geopoetics* (Dingwall: Sandstone, 2007).

Menorca, 'The North Minorcan Marine Reserve', <http://www.menorca.es/contingut.aspx?IDIOMA=3&idpub=8863> [accessed 10 April 2020].

Merleau-Ponty, Maurice, *Phenomenology of Perception*, trans. by Colin Smith (London: Routledge and Kegan Paul, 1962), trans. rev. by Forrest Williams (1981; repr. 2002).

Mitchell, W. J. T., 'Imperial Landscape', in *Landscapes and Power*, ed. by W. J. T. Mitchell (Chicago, IL: University of Chicago Press, 2002), 7–36.

Modeen, Mary, 'Breaking the Boundaries of "Self": Representations of Spatial Indeterminacy', *Architecture and Culture: Transgression: Body and Space*, 2 (2014), 335–358.

———, 'Love from a Distance', Paper Presented at *Catchment/PLaCE Mapping Spectral Traces*, Bristol, United Kingdom, 2011.

Modood, Tariq, *Multiculturalism: A Civic Idea* (London: Polity Press, 2007).

Montiel, Anya, 'Reclaiming the Landscape: The Art of Lewis deSoto', *American Indian*, 13 (2012), 24–30.

Morphy, Howard, *Aboriginal Art* (London: Phaidon, 1998).

Morton, Timothy, *The Ecological Thought* (Cambridge, MA: Harvard University Press, 2010).

Mullen, Lincoln, 'Deep Maps', *Spatial Humanities Workshop* (2015), <http://lincolnmullen.com/projects/spatial-workshop/deep-maps.html> [accessed 10 April 2020].

Mulligan, Martin and Stuart Hill, eds, *Ecological Pioneers: A Social History of Australian Ecological Thought and Action* (Cambridge, MA: Cambridge University Press, 2001).

Naess, Arne, *The Ecology of Wisdom: Writings by Arne Naess*, ed. by Alan Drengson and Bill Devall (Berkeley, CA: Counterpoint, 2008).

Nancy, Jean-Luc, *The Inoperative Community*, trans. by Peter Connor and Others (Minneapolis: University of Minnesota Press, 1991).

Napier, A. David, *Foreign Bodies: Performance, Art, and Symbolic Anthropology* (Oakland: University of California Press, 1992).

———, *Making Things Better: A Workbook on Ritual, Cultural Values, and Environmental Behavior* (Oxford: Oxford University Press, 2014).

———, *Masks, Transformation, and Paradox* (Berkeley: University of California Press, 1992).

National Exhibitions Touring Support Victoria, 'The Stony Rises Project', <https://netsvictoria.org.au/exhibition/the-stony-rises-project/> [accessed 10 April 2020].

Nicholson, Geoff, *The Lost Art of Walking: The History, Science, Philosophy, Literature, Theory and Practice of Pedestrianism* (Essex: Harbour Books (East), 2011).

Nietzsche, Friedrich, *Twilight of the Idols, or, How to Philosophize with a Hammer* (1889), English Version, trans. by R.J. Hollingdale (New York: Penguin, 2003).

Oelschlager, Max, *The Wilderness Condition: Essays on Environment and Civilisation* (Washington, DC: Island Press, 1992).

Paul, Kalpita Bhar, 'The Import of Heidegger's Philosophy into Environmental Ethics: A Review', *Ethics and the Environment*, 22, no. 2 (2017), 79–98, <http://www.jstor.org/stable/10.2979/ethicsenviro. 22.2.04> [accessed 10 April 2020].

Paulson, Graham, 'Aboriginal Spirituality', *Australians Together* [n.d.], <https://www.australianstogether.org.au/discover/indigenous-culture/aboriginal-spirituality/> [accessed 10 April 2020].

Pearson, Michael, *'In Comes I': Performance, Memory and Landscape* (Exeter: University of Exeter Press, 2006).

Pearson, Mike, and Michael Shanks, *Theatre/Archaeology: Disciplinary Dialogues* (London: Routledge, 2001).

Petrini, Carlo, *Slow Food: The Case for Taste* (New York: Columbia University Press, 2001).

Phillips, Graham, 'Secrets of the Stones', *Sydney Morning Herald* (13 March 2003), <https://www.smh.com.au/national/secrets-of-the-stones-20030313-gdgf3f.html> [accessed 10 April 2020].

Phillips, Perdita, *The Sixth Shore 2009–2014*, <http://www.perditaphillips.com/-current- projects/the-sixth-shore/> [accessed 10 April 2020].

Pignarre, Philippe, and Isabelle Stengers, *Capitalist Sorcery: Breaking the Spell*, trans. and ed. by Andrew Goffey (Basingstoke: Palgrave Macmillan, 2011).

Pinker, Steven, 'Stranger Than Fiction', *Guardian Review* (13 August 2016), 3.

Portway, Joshua, and Lise Autogena, *Black Shoals Stock Market Planetarium*, 2000–Present, Installation, <http://www.blackshoals.net/> [accessed 10 April 2020].

Ransley, Tim, and Andrew Feitz, 'Navigating Australia's Largest Groundwater Resource', *Australian Government: Geoscience Australia*, <https://www.ga.gov.au/news-events/features/navigating-australias-largest- groundwater-resource> [accessed 10 April 2020].

Ratti, Carlo, and Others, 'The Power of Networks: Beyond Critical Regionalism', *The Architectural Review* (23 July 2013), <https://www.architectural-review.com/essays/the-power-of-networks-beyond-critical-regionalism/8651014.article> [accessed 10 April 2020].

Read, Simon, 'The Power of the Ooze', in *The Power of the Sea: Making Waves in British Art 1790–2014*, ed. by Janette Kerr and Christiana Payne (Bristol: Sansom & Company, 2014), 45–54.

Reiter, Bernd, ed., *Constructing the Pluriverse: The Geopolitics of Knowledge* (Durham, NC: Duke University Press, 2018).

Relph, Edward, *Place and Placelessness* (London: SAGE Publications, 2008).

Richardson, Laurel, 'The Consequences of Poetic Representation: Writing the Other, Rewriting the Self', in *Investigating Subjectivity: Research on Lived Experience*, ed. by Carolyn Ellis and Michael Flaherty (Los Angeles, CA: SAGE Publications, 1992), 125–137.

Ricoeur, Paul, *The Course of Recognition*, trans. by David Pellauer (London: Harvard University Press, 2005).

——, *Memory, History, and Forgetting*, trans. by Kathleen McLaughlin and David Pellauer (Chicago, IL: University of Chicago Press, 2004).

——, *Time and Narrative (Temps et Récit)*, trans. by Kathleen McLaughlin and David Pellauer, 3 vols (Chicago, IL: University of Chicago Press, 1983–85).

Rilke, Rainer Maria, *The Duino Elegies*, trans. By Vita Sackville West (London: Hogarth Press, 1931).

Roberts, Les, ed., *Deep Mapping* (Basel: MDPI AG – Multidisciplinary Digital Publishing Institute, 2016).

——, 'Preface: Deep Mapping and Spatial Anthropology', in *Deep Mapping*, ed. by Les Roberts (Basel: MDPI AG – Multidisciplinary Digital Publishing Institute, 2016), VII–XV.

——, 'The Rhythm of Non-Places: Marooning the Non-Self in Depthless Space', in *Deep Mapping*, ed. by Les Roberts (Basel: MDPI AG – Multidisciplinary Digital Publishing Institute, 2016), 155–185.

Robin, Libby, 'Histories for Changing Times: Entering the Anthropocene?', *Australian Historical Studies*, 44, no. 3 (2013), 329–340.

Rorty, Richard, *The Mirror of Nature* (Princeton, NJ: Princeton University Press, 1979).

Rose, Mitch, 'Marking a Life', in *The Creative Critics: Writing as/about Practice*, ed. by Katja Hilevaara and Emily Orley (London: Routledge, 2018), 200–205.

Said, Edward, 'Invention, Memory and Place', *Critical Inquiry*, 26, no. 2 (Chicago, IL: University of Chicago Press, 2002), 241–260.

Saltelli, Andrea, and Mario Giampieto, 'What Is Wrong with Evidence Based Policy, and How Can it be Improved?', *Futures*, 91 (2017), 62–71.

Scarry, Elaine, *On Beauty and Being Just* (Princeton, NJ: Princeton University Press, 1999).

Schiavini, Cinzia, 'Writing the Land: Horizontality, Verticality and Deep Travel,' in *PrairyErth: A Deep Map*, by William Least Heat-Moon (New York: Houghton Mifflin, 1999), 79–98.

Schumacher, E. F., *Small Is Beautiful: A Study of Economics as if People Mattered* (London: Blond and Briggs, 1975).

Scottish Centre for Geopoetics (2020), <http://www.geopoetics.org.uk/> [accessed 10 April 2020].

Sebald, W. G., *Austerlitz*, trans. by Anthea Bell (New York: Random House, 2001).

Sennett, Richard, *The Craftsman* (Cambridge, MA: Yale University Press, 2008; New York: Penguin, 2009).

Shotter, John, 'Life Inside Dialogically Structured Mentalities: Bakhtin's and Voloshinov's Account of Our Mental Activities as Out in the World Between Us', in *The Plural Self: Multiplicity in Everyday Life*, ed. by John Rowan and Mick Cooper (London: SAGE Publications, 1999), 71–92.

Siegenthaler, David, 'Earth Walk: A Deep Ecology Perspective and Critique of the Mainstream Environmental Movement', *Unbound* (2012), <https://justiceunbound.org/earth-walk/> [accessed 9 November 2013].

Smith, Phil, *Mythogeography: A Guide to Walking Sideways* (Charmouth: Triarchy Press, 2010).

Smithson, Robert, *Robert Smithson: The Collected Writings*, ed. by Jack Flam, 2nd edn (Berkeley: University of California Press, 1996).

Snyder, Gary, 'The Politics of Ethnopoetics', Paper based on a Talk Presented at the *Ethnopoetics* Conference at the University of Wisconsin (April 1975), <http://angg.twu.net/LATEX/poep.pdf> [accessed 24 November 2017].

Solnit, Rebecca, *The Faraway Nearby* (London: Granta, 2013).

———, *Wanderlust: A History of Walking* (London: Verso, 2000).

Sontag, Susan, *Regarding the Pain of Others* (New York: Picador/Farrar, Straus and Giroux, 2003).

———, *The Volcano Lover* (New York: Farrar, Straus and Giroux, 1992).

Spolsky, Ellen, *Gaps in Nature: Literary Interpretation and the Modular Mind* (Albany: SUNY Press, 1993).

Stegner, Wallace, *The Sense of Place*: A *Nine Page Pamphlet* (Madison: Wisconsin Humanities Committee, 1986).

———, *Wolf Willow*: A *History, a Story, and a Memory of Silent Plains Frontier* (New York: Penguin Publishing, 2000).

Stephens, Chris, *Peter Lanyon: At the Edge of Landscape* (London: 21 Publishing, 2000).

The Stony Rises Project, 'The Works', <http://thestonyrisesproject.com/the-exhibition/the-works/> [accessed 16 November 2018].

Strand, Mark, 'Keeping Things Whole', in *Selected Poems* (New York: Knopf, 1990).

Szewcyk, Monika, 'The Art of Conversation Part I', *e-flux*, 3 (2009), <http://www.e-flux.com/journal/art-of-conversation-part-I> [accessed 10 April 2020].

Terminalia Festival, 'Terminalia – Festival of Psychogeography', <http://terminaliafestival.org/> [accessed 10 April 2020].

Thompson, Jon, 'Campus Camp', in *Artists in the 1990s: Their Education and Values*, ed. by Paul Hetherington (London: Wimbledon School of Art in Association with Tate Gallery, 1994), 44–48.

Thoreau, Henry David, *Walden: Or, Life in the Woods* (1854), (Princeton: Princeton University Press, 2004).

Tiampo, Ming, and Alexandra Munroe, 'Gutai: Splendid Playground February 15–May 8, 2013', *Guggenheim*, <http://web.guggenheim.org/exhibitions/gutai/> [accessed 10 April 2020].

Till, Karen, *Mapping Spectral Traces: Exhibition Publication* (Blacksburg: Virginia Tech College of Architecture and Urban Studies, 2010).

Tiravanija, Rirkrit, 'Night School', New Museum Seminar, New York (25 September 2010). 'The Land' Project.

Turner, Cathy, 'Palimpsest or Potential Space? Finding a Vocabulary for Site-Specific Performance', *New Theatre Quarterly*, 20, no. 4 (2004), 373–390.

Turner, Dale, *This Is Not a Peace Pipe: Towards a Critical Indigenous Philosophy* (Toronto: University of Toronto Press, 2006).

Turner, Victor, *The Ritual Process: Structure and Anti-Structure* (Chicago, IL: Aldine, 1969).

USA National Endowment for the Arts, 'Creative Placemaking Guidelines and Report Launched' (21 May 2015), <https://www.arts.gov/news/2015/creative-placemaking-guidelines-and-report-launched> [accessed 10 April 2020].

Vanishing of the Bees, dir. by George Langworthy and Maryam Henein (Hive Mentality Films and Hipfuel Films, 2009).

Veal, Clare, 'Bringing the Land Foundation Back to Earth: A New Model for the Critical Analysis of Relational Art', *Journal of Aesthetics & Culture*, 6, no. 1 (2014), 178–183, <https://doi.org/10.3402/jac.v6.23701>.

Vlachaki, Elika, 'Objects as Portals: The Kapodistrian Orphanage and Aegina Prison' (unpublished doctoral dissertation, University of Dundee, 2019).

Walters, Victoria, *Joseph Beuys and the Celtic Wor(l)d: A Language of Healing* (London: Lit Verlag, 2012).

Ward, Ossian, *Ways of Looking: How to Experience Contemporary Art* (London: Laurence King, 2014).

Warf, Barney, 'Deep Mapping and Neogeography', in *Deep Maps and Spatial Narratives*, ed. by David Bodenhamer, John Corrigan, and Trevor Harris (Bloomington: Indiana University Press, 2015), 134–149.

Watkins, Mary, '"Breaking the Vessels": Archetypal Psychology and the Restoration of Culture, Community and Ecology', in *Archetypal Psychologies: Reflections in Honor of James Hillman*, ed. by Stanton Marlan (New Orleans, LA: Spring Journal Books, 2008), 415–438.

———, 'From Hospitality to Mutual Accompaniment: Addressing Soul Loss in the Citizen-Neighbour', in *Borders and Debordering: Topologies, Praxes, Hospitableness*, ed. by Tomaž Gruovnik, Eduardo Mendieta and Lenart Škof (London: Lexington Books, 2018), 25–39.

Watkins, Mary, and Helene Shulman, *Towards Psychologies of Liberation* (London: Palgrave Macmillan, 2008).

White, Kenneth, *Geopoetics: Place, Culture, World* (Glasgow: Alba Books, 2003).

———, 'The International Institute of Geopoetics: Inaugural Text', *Scottish Centre for Geopoetics* (1989) <http://www.geopoetics.org.uk/what-is-geopoetics/> [accessed 24 November 2017].

———, *The Tribal Dharma: An Essay on the Work of Gary Snyder* (Dyfed: Unicorn Press, 1975).

White, Melissa and Glen Scholz, *Prioritising Springs of Ecological Significance in the Flinders Ranges* (Adelaide: Department of Water, Land and Biodiversity Conservation, 2008).

Whiteman, Honor, 'How Does Lack of Sleep Impair Memory Foundation?', *Medical News Today* (10 April 2017), <http://www.medicalnewstoday.com/articles/316863.php> [accessed 10 April 2020].

Williams, Leanne M., and Others, 'Amygdala–Prefrontal Dissociation of Subliminal and Supraliminal Fear', *Human Brain Mapping*, 27, no. 8 (2006), 652–661.

Wolf, Fred, A., 'The Dreamtime' (Chapter 9), in *The Dreaming Universe: A Mind-Expanding Journey into the Realm Where Psyche and Physics Meet* (New York: Simon & Schuster, 1994).

Wood, Denis, *Everything Sings: Maps for a Narrative Atlas* (Los Angeles, CA: Siglio, 2011).

———, 'Mapping Deeply', in *Deep Mapping*, ed. by Les Roberts (Basel: MDPI AG – Multidisciplinary Digital Publishing Institute, 2016), 15–29.

Worthy, Martin, 'Warp and Weft on the Loom of Lat/Long', in *Deep Maps and Spatial Narratives*, ed. by David Bodenhamer, John Corrigan, and Trevor Harris (Bloomington: Indiana University Press, 2015), 203–222.

Wright, James. A., 'The Jewel', in *The Branch Will Not Break* (Middletown, CT: Wesleyan University Press, 1963), 17.

Wrights & Sights (2018), 'A Manifesto for a New Walking Culture', <http://www.mis-guide.com> [accessed 10 April 2020].

Young-Bruehl, Elizabeth, *Hannah Arendt: For Love of the World* (New Haven, CT: Yale University Press, 1982).

Zitzewitz, Karin, 'Past Futures of Old Media: Gulammohammed Sheikh's Kaavad: Travelling Shrine: Home', in *Media and Utopia: History, Imagination, Technology*, ed. by Arvind Rajagopal and Anupama Rao (New Delhi: Routledge, 2016), 189–208.

Index

Note: *Italic* page numbers refer to figures and page numbers followed by "n" denote endnotes.